Size Reduction of Divided Solids

There are no such things as applied sciences,
only applications of science.
Louis Pasteur (11 September 1871)

Dedicated to my wife, Anne, without whose unwavering support, none of this
would have been possible.

Industrial Equipment for Chemical Engineering Set

coordinated by
Jean-Paul Duroudier

Size Reduction of Divided Solids

Jean-Paul Duroudier

ELSEVIER

First published 2016 in Great Britain and the United States by ISTE Press Ltd and Elsevier Ltd

ISTE Press Ltd
27-37 St George's Road
London SW19 4EU
UK

www.iste.co.uk

Elsevier Ltd
The Boulevard, Langford Lane
Kidlington, Oxford, OX5 1GB
UK

www.elsevier.com

Notices

For information on all our publications visit our website at http://store.elsevier.com/

British Library Cataloguing-in-Publication Data
A CIP record for this book is available from the British Library
Library of Congress Cataloging in Publication Data
A catalog record for this book is available from the Library of Congress
ISBN 978-1-78548-185-7

Printed and bound in the UK and US

Contents

Preface

The observation is often made that, in creating a chemical installation, the time spent on the recipient where the reaction takes place (the reactor) accounts for no more than 5% of the total time spent on the project. This series of books deals with the remaining 95% (with the exception of oil-fired furnaces).

It is conceivable that humans will never understand all the truths of the world. What is certain, though, is that we can and indeed must understand what we and other humans have done and created, and, in particular, the tools we have designed.

Even two thousand years ago, the saying existed: "faber fit fabricando", which, loosely translated, means: *"c'est en forgeant que l'on devient forgeron"* (a popular French adage: *one becomes a smith by smithing*), or, still more freely translated into English, "practice makes perfect". The "artisan" (faber) of the 21st Century is really the engineer who devises or describes models of thought. It is precisely that which this series of books investigates, the author having long combined industrial practice and reflection about world research.

Scientific and technical research in the 20th century was characterized by a veritable explosion of results. Undeniably, some of the techniques discussed herein date back a very long way (for instance, the mixture of water and ethanol has been being distilled for over a millennium). Today, though, computers are needed to simulate the operation of the atmospheric distillation column of an oil refinery. The laws used may be simple statistical correlations but, sometimes, simple reasoning is enough to account for a phenomenon.

Since our very beginnings on this planet, humans have had to deal with the four primordial "elements" as they were known in the ancient world: earth, water, air and fire (and a fifth: aether). Today, we speak of gases, liquids, minerals and vegetables, and finally energy.

The unit operation expressing the behavior of matter are described in thirteen volumes.

It would be pointless, as popular wisdom has it, to try to "reinvent the wheel" – i.e. go through prior results. Indeed, we well know that all human reflection is based on memory, and it has been said for centuries that every generation is standing on the shoulders of the previous one.

Therefore, exploiting numerous references taken from all over the world, this series of books describes the operation, the advantages, the drawbacks and, especially, the choices needing to be made for the various pieces of equipment used in tens of elementary operations in industry. It presents simple calculations but also sophisticated logics which will help businesses avoid lengthy and costly testing and trial-and-error.

Herein, readers will find the methods needed for the understanding the machinery, even if, sometimes, we must not shy away from complicated calculations. Fortunately, engineers are trained in computer science, and highly-accurate machines are available on the market, which enables the operator or designer to, themselves, build the programs they need. Indeed, we have to be careful in using commercial programs with obscure internal logic which are not necessarily well suited to the problem at hand.

The copies of all the publications used in this book were provided by the *Institut National d'Information Scientifique et Technique* at Vandœuvre-lès-Nancy.

The books published in France can be consulted at the *Bibliothèque Nationale de France*; those from elsewhere are available at the British Library in London.

In the in-chapter bibliographies, the name of the author is specified so as to give each researcher his/her due. By consulting these works, readers may gain more in-depth knowledge about each subject if he/she so desires. In a reflection of today's multilingual world, the references to which this series points are in German, French and English.

The problems of optimization of costs have not been touched upon. However, when armed with a good knowledge of the devices' operating parameters, there is no problem with using the method of steepest descent so as to minimize the sum of the investment and operating expenditure.

Grinding: Principles and Theories

1.1. Grinding

1.1.1. *Introduction*

Chemistry uses bulk solids with average particle sizes ranging from 1 mm to 100 μm (fine), including particle sizes less than 20 μm (ultrafines).

For fibrous or plastic materials, splitting or shredding is used. However, for classic bulk solids (minerals, crystals, etc.), pressure or collisions are used and shearing is used less often.

Grinding equipment includes:

– a storage hopper for the bulk solid to be processed;

– a feed system using a screw conveyor, a conveyor belt or a vibrating conveyor;

– a comminution device;

– a classifier which generally uses a centrifugal force balanced by air (or water) friction;

– air decontamination equipment (usually dust extraction) and a fan for its removal.

Hixon [HIX 91] reviews different possible processes.

The wearing down of grinding surfaces requires part replacements (for example, plates and millstones).

The shape of the output obtained is important. As such, the gravel's shape must be compact, that is, its three dimensions must be similar.

The average size of pigmented powders determines its color.

Wet grinding (after dispersion in a liquid) could lead to the flocculation of particles smaller than 10 μm.

Humidity greater than 50% (on-wet) liberates grains and improves their grinding.

The impact of temperature could be eliminated in the following ways, if it is too high:

– mix crushed ice into the feed system;

– cool the device with liquid nitrogen.

The risk of explosion could be eliminated by operating in a neutral atmosphere (CO_2, N_2, washed combustion gases).

1.1.2. *Grain size*

The size of the feed has the following order of magnitude:

crushing	250 mm
first-stage grinding	75 mm
second-stage grinding	25 mm
gas jet mill	0.1 mm

The reduction ratio is:

$$R = \frac{\text{average size of feed}}{\text{average size of the exiting output}}$$

The ratio R has a magnitude of 6 for a crusher and could reach up to 400 for a fluid energy mill.

The *specifications* imposed on the mill often have two limits. For example:

– there must be more than 80% of the mass that is less than 0.3 mm;

– there must be less than 5% of the mass that is less than 0.05 mm.

The fines produced (dry dust and wet pulp viscosity) are:

– very large with grinding track mills;

– much less with roller mills;

– non-existent with shredders (sugar cane) and splitting equipment.

1.1.3. *Output granulometry*

We will often call the "granulometry" of a bulk solid the distribution of particle sizes by mass.

If m_i is the mass remaining on the screen i, we can study the variation in m_i in comparison with its size x_j, which is the *frequency distribution*. The *cumulative distribution,* versus x_j, looks at variations in the underflow P (see section 1.4.1):

$$P = \sum_{j+1}^{n} m_i$$

The average size of the output obtained from a comminution process is linked to the type of the process used.

Primary crushing	$5 \text{ cm} < \overline{x} < 10 \text{ cm}$
Secondary crushing	$1 \text{ cm} < \overline{x} < 5 \text{ cm}$
Tertiary crushing	$1 \text{ mm} < \overline{x} < 1 \text{ cm}$
Coarse grinding	$100 \text{ μm} < \overline{x} < 1 \text{ mm}$
Fine grinding	$20 \text{ μm} < \overline{x} < 100 \text{ μm}$
Ultrafine grinding	$0.1 \text{ μm} < \overline{x} < 20 \text{ μm}$

Table 1.1. *Average size of output*

Granulometry of the output obtained could be more or less spread out. We refer to:

– granules, as being the largest particles (which are characterized as grains);

– fines, as being the smallest particles (which are characterized as powder).

Basically:

– an excessive amount of fines reflects ineffective over-grinding and energy wastage;

– fines generate dust;

– they may, by a cushion effect, reduce grinding efficiency (ball and track mills);

– in the presence of a little humidity, the output becomes sticky which may cause machinery to "jam".

The size of "fines" is less than 20 μm ("fines" become ultrafines).

When grinding a load, the granulometry of the load develops over time.

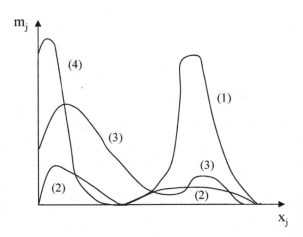

Figure 1.1. *Progression of the granulometry (curve 1) of a product during grinding*

At the beginning (Figure 1.1), product size distribution is unimodal (curve 1). Then, it becomes bimodal (curves 2 and 3) and, finally, it becomes unimodal again (curve 4).

The fragments obtained from breaking grains are generally considerably much smaller than the grains being milled, which explains the position of the curve (4).

If variations in $\log_{10}m_j$ are represented in comparison with $\log_{10}x_j$, we obtain the curve shown in Figure 1.2.

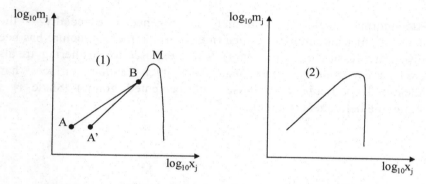

Figure 1.2. *Granulometry of a product coming out of a ball mill*

The frequency curve generally has a straight line portion AB and a "hump" at M. Sizes at this hump correspond to grains that are too large to fit between:

– two balls in a ball mill;

– two cylinders in a roller mill.

Rod mills are superior at breaking down the largest grains.

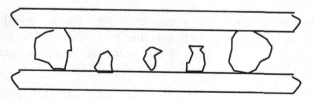

Figure 1.3. *Striking grains with two rods*

The rods hit the large grains while the smallest grains are not touched. The result is that the output exiting the rod mill (curve (2) of Figure 1.2) does not have the hump M of curve (1).

In addition, note that rods with a diameter of 2 cm could hit grains that 2 cm balls are unable to hit.

A steep slope for the line A'B shows less fines than in the line AB.

1.1.4. *Heterogeneous rocks*

Metamorphic rocks are extremely heterogeneous, especially if they contain a significant amount of mica (metamorphic rocks structure has been modified by pressure or temperature). Rocks exposed to weathering are also heterogeneous from the fact that their inter-granular surfaces have disintegrated (particularly in those with cemented grains). Shale is an example of a heterogeneous rock.

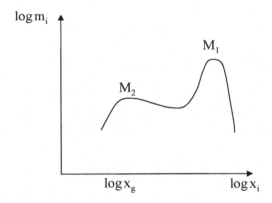

Figure 1.4. *Granulometry of a heterogeneous rock*

In the fines region, we see a maximum M_2 which corresponds to an average grain size $\bar{x}$ of rocks held together by a "cement" or pressure. The more the number of large grains in the rock, the more the maximum M_2 shifts to the right.

If all the grains in the rock have a similar size, then the maximum M_2 is very sharp. Some rocks display 2 or 3 maxima of the type M_2.

Metamorphic rocks are very heterogeneous (gneisses).

Grain boundaries are weak zones. Wind and water erode heterogeneous rocks as grain boundaries are quickly broken down.

If a rock is permeable to moisture, it will be sensitive to freezing.

Some grains in rocks are more difficult to break. The result is 2 or 3 maxima in variations of m_j function of x_j (see Figure 1.4).

1.2. Empirical laws governing overflow and underflow of a bulk solid

1.2.1. Schumann's Simple Law [SCH 40] for underflow (see section 1.4.1)

The underflow P(x) is represented as a function of the particle size x by the function:

$$P(x) = \left(\frac{x}{x_o} \right)^{\alpha}$$

α: module distribution (dimensionless)

x_o: module size (m).

In double logarithmic coordinates, this correlation is shown as a straight line. Crabtree *et al.* [CRA 64] characterized three fracture modes during a grinding process:

– collision;

– surface abrasion;

– emuscation of edges and tip breaking.

In fragmentation by collision, the distribution module is very close to 1. However, combining abrasion with emuscation provides a distribution module α that is clearly less than 1. Combining three fracture mechanisms no longer produces a straight line but rather a curve (with double logarithmic coordinates) that is concave up. For small particle sizes, the slope of the curve is much less than 1. In fact, surface abrasion, emuscation of edges and tip breaking create small particles that were torn from mother particles.

1.2.2. *Frequency distribution: according to Schumann's Law [SCH 40]*

From Schumann's Law [SCH 40], the frequency distribution is obtained by differentiation:

$$f(r) = \frac{dP}{dx} = \frac{\alpha}{x_o^\alpha} x^{\alpha-1}$$

In other words:

$$Lnf = (\alpha - 1)Lnx + Ln\left(\frac{\alpha}{x_o^\alpha}\right)$$

Or even:

$$Ln\, m_i = kLnx_i + LnC$$

1.2.3. *The Rosin–Rammler formula for overflow*

Rosin and Rammler [ROS 34] proposed a simple empirical expression for the overflow R as a function of the particle diameter x:

$$R = \varrho^{-bx^n}$$

Of course, the underflow P is:

$$P = 1 - R$$

According to the authors (and in general), expressing R explains what happens when all particles fall inside the range of 1–500 μm.

Knowing that overflows R_1 and R_2 correspond to particle diameters x_1 and x_2, we can get the value of the two parameters n and b:

$$n = \frac{Ln\left(Ln\dfrac{1}{R_1}\right) - Ln\left(Ln\dfrac{1}{R_2}\right)}{Lnx_1 - Lnx_2}$$

$$Lnb = Ln\left(Ln\frac{1}{R_1}\right) - nLnx_1$$

NOTE.–

Let us recall the Gaudin–Meloy [GAU 62] formula:

$$R = \left(1 - \frac{x}{x_o}\right)^r$$

These authors provide perspectives about fractures in a volume, on a surface or an edge.

1.2.4. *The Harris distribution [HAR 69]*

While the Schumann [SCH 40] and Rosin–Rammler [ROS 34] distributions provide only two parameters, Harris [HAR 69] proposed a three-parameter equation, which allows for a much closer rendering of results from grinding.

According to Harris, the underflow with a dimension x is given by:

$$P(2) = \left[1 - \left(\frac{x}{x_o}\right)^s\right]^r$$

Barbery [BAR 71] uses a regression method to determine the value of Harris' parameters x_o, s, r.

NOTE.–

Harris [HAR 69] proposes a representation system that includes:

– an anamorphosis of P(x) on the ordinate;

– a logarithmic scale for x/x_o on the abscissa.

In this way (refer to author's Figure 1), the function P(x) is represented by a straight line on a large part of the variation of this function.

1.2.5. *Log-normal distribution*

Epstein [EPS 48] explains that, if each grinding event splits a particle into two equal parts, the result would be a log-normal size distribution after a dozen or so events.

For a normal distribution, the frequency of the size x is:

$$F(x)dx = \frac{\exp(-x^2/2)}{\sqrt{2\pi}}dx$$

For a log-normal distribution, we obtain the variable: $\dfrac{\text{Ln } x}{\sigma}$

$$F_L(x)dx = \frac{1}{\sigma\sqrt{2\pi}}\exp\left(-\frac{(Lnx)^2}{2\sigma^2}\right)d\,Ln\,x$$

The total underflow is then:

$$P(x) = \int_o^x F_L(x)dx = \text{erf}\left(\frac{Ln\,x}{\sigma}\right)$$

This distribution is only used for the diameter of drops emerging from a crusher and is in no way applicable to grinding processes, except by Austin and Bagga [AUS 81a] for the selection function.

NOTE.–

Spiegel [SPI 92] gives values to the integral that we called $P(x)$.

NOTE.–

With concern to homogeneity, we should have introduced Ln σ instead of σ.

The log-normal distribution law can also be obtained from the Gauss normal distribution law:

$$g(y)dy = \frac{1}{\sigma\sqrt{2\pi}}\exp-\frac{(y-y_{mo})^2}{2\sigma^2}dy$$

Where:

$$y = \ln t \quad \text{and} \quad dy = \frac{dt}{t}$$

$$f(t)dt = \frac{dt}{\sigma t\sqrt{2\pi}}\exp-\left[\frac{\left(Ln\dfrac{t}{t_{mo}}\right)^2}{2\sigma^2}\right]$$

t_{mo}: average value of t

When t goes to zero, we also see $f(t)$ going to zero.

Of course, in a more empirical manner, we write:

$$f(t)dt = \frac{dt}{\sigma\sqrt{2\pi}}\exp-\left[\frac{\left(Ln\dfrac{t}{t_{mo}}\right)^2}{2\sigma^2}\right]$$

These two expressions are practically not used.

1.3. Physics of fragmentation

1.3.1. *Cracking phenomenon [SCH 72]*

Let us look at a sample with section Ω and subjected to stress with a uniform traction σ_0, the initial length of the sample is L and the material's modulus of elasticity is given by E.

The elastic energy stored in the sample is:

$$U = \Omega \int_0^{\sigma_0} dL$$

Where:

$$dL = \frac{L}{E} d\tau$$

Therefore:

$$U = \frac{\Omega L}{E} \int_0^{\sigma_0} \sigma \, d\sigma = \frac{\sigma_0^2 \Omega L}{2E} = \frac{\sigma_0^2 V}{2}$$

Let us now look at the existing crack in the sample. It can be compared with a flat pocket stitched on its perimeter except along its opening. If A is the fresh new surface created by the crack, the surface of each of its sides is A/2. We will call crack front the seam on our pocket. This front is defined not by the surface but by the perpendicular line in the plane of Figure 1.5.

We established in the theory of elasticity the stress from a crack with a straight opening of length 2a, this opening being perpendicular to a uniform traction stress τ_0 existing in the material. Figure 1.5 shows the crack front.

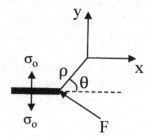

Figure 1.5. *Crack front F perpendicular to the plane of the figure*

By selecting polar coordinates in the plane perpendicular to the front, the coordinate origins are exactly on the front. We show that:

$$\sigma_x = \sigma_o \sqrt{a / 2\rho} f_x (\theta)$$

$$\sigma_y = \sigma_o \sqrt{a / 2\rho} f_y (\theta)$$

$$\sigma_{xy} = \sigma_o \sqrt{a / 2\rho} f_{xy} (\theta)$$

These stresses contribute to the sample's stored energy. When the crack gets bigger, its front is displaced and sweeps the surface dA/2. The theory of elasticity shows that elastic energy used by the propagation of the crack is:

$$-dU = G\frac{dA}{2} \quad \left(G \text{ is measured in J.m}^{-2} \text{ or N.m}^{-1}\right)$$

In this expression:

$$G = K\frac{\sigma_o^2 a}{E}$$

The coefficient K depends on the shape of the crack.

According to the intensity of the stress σ_o, the front will be displaced at a speed of $v_F = da/d\tau$, which increases with G, as shown in Figure 3 in Schoenert's [SCH 72] publication. If G is zero, the speed v_F will also be zero and, if G is very large, the speed will approach a limit that is equal to half or one-third the speed of sound in a given solid. The possible range of variations in v_F is between 0.01 m.s^{-1} and 300 m.s^{-1}.

The time taken to apply stress to a grain is τ_A. For a crack to travel throughout a grain of diameter d_g, there must be:

$$v_F > d_g / \tau_A$$

Therefore, in a ball mill, the duration of the collision of a ball on a grain is $\tau_A = 0.001$ s and, if we let $d_g = 0.01$ m, we should get:

$$v_F > 10\,\text{m.s}^{-1}$$

This minimum speed corresponds to a minimum value of G that we often call 2β. Practical values of β are as follows:

Material	β (J.m^{-2})
Glass	1–10
Plastic	10–10^3
Metal	10^3–10^5

Table 1.2. *Minimum values for G*

Note that the fracture energy of solids is of the order of 0.1–1 J.m^{-2}. For solids this magnitude is equivalent to a liquid surface tension. (1 J.m^{-2} = 1 N.m^{-1}). The surface energy is weak in comparison with β and explains derisory theoretical yields given by early comminution theorists.

At the crack front, energy is used for plastic deformation, which creates heat emission and often a substantial increase in temperature that can be measured with a thermocouple. This deformation is rapid (typically 10^{-6} s) and the resulting structure is literally "frozen" after the front passes through. The result is that during the grinding of a crystallized substance, the fragment surfaces could become more or less amorphous. This is the case with sugar.

Plastic deformation, although rapid, is not instantaneous and, if the speed of the front is increased, there will be less time available for deformation, and therefore less energy is used up.

Inversely, let us assume that stresses are introduced slowly and that the level of prior stresses is high. So it will be in a significant volume around the front and, with the elastic limit being achieved in this extended region, the energy used for plastic deformation at the crack front will be high.

In practice, the front starts slowly and then speeds up its progression. In fact, G is proportional to a, but, if the available elastic energy is used up or very low, the crack will stop spreading.

For a crack to begin spreading, G, hence a, must not be zero, which is postulated by pre-existing defects in the material (dislocations in the case of crystals). If these defects are very localized (a is small), a significant amount of stress σ_0 must be applied to cause the crack to grow and finally there is to be a fracture.

Note that the effect of temperature on cracking is important. The speed of the front could be multiplied by 10 when the temperature is increased from 20 to 80°C (Figure 3 of Schoenert's publication).

1.3.2. *Elastic behavior and plastic behavior*

If a sphere is compressed between two planes, a conical zone develops at the two points of contact. On the inside of this zone compression is high.

This cone tends to sink in like a corner on the inside of the sphere. Significant stress from tension appears on the periphery of the cone and the sphere breaks up in layers like onion skin. The cones are finely fragmented and create fines.

Inversely, a sphere with plastic behavior breaks in segments like an orange without leading to the formation of fines.

1.3.3. *Distinction between slow crushing and collision*

Slow crushing creates elastic stresses where a good proportion of it does not lead to cracks and energy is dispersed into heat. The paper by Kanda *et al.* [KAN 86] gives the expression for the specific breaking energy from Hertz' theory for a sphere compressed between two plane surfaces.

A collision generates elastic waves that reflect against the particle's walls and interfere between themselves. Locally, a maximum of stress could generate a crack from a defect but the principal part of the wave energy is absorbed and dissipated as heat.

In the two cases (collision and compression damping), the created fragments escape at high speeds of the order of 300–1,500 m.s^{-1} and, then, their kinetic energy is dissipated as heat through friction with the surroundings.

1.3.4. *Critical diameter*

Grains with a size greater than approximately 100 μm break where defects are most extensive and, in the resultant fragments, the defects that remain will have the lowest effectiveness in terms of fracture. As such, there will be less crack initiation with a greater decrease in particle size. Such a grain will therefore divide into a smaller number of fragments. For a given critical size, the grain will no longer split and will have entirely plastic behavior.

Material	d_c (μm)
Calcite	0.1
Glass	0.5
Cement	0.5–1
Quartz	1
Limestone	3–5
Coal	35

Table 1.3. *Critical size of selected rocks and minerals*

For moderately hard crystals such as organic crystals and many salts, d_c could get to a few tens of micrometers.

Experimentally, a sphere of size 100 μm or greater could have several cracks on compression and the stress–strain curve will have saw-tooth irregularities. Inversely, the same curve will have a consistent appearance if the size of the sphere is less than the critical size.

An additional argument justifies the plastic behavior of fine grains. In fact, elastic energy stored in a particle decreases as the cube of its diameter. Yet, the energy required to break the particle, proportional to the newly formed surface, only decreases as the square of its diameter. It is therefore difficult to accumulate enough energy in a fine particle for it to break.

Griffith [GRI 20] showed that resistance to the fracture of a glass thread is 20 times greater when the diameter of the thread goes from 1 mm to 100 µm.

Therefore, if the size of the particle is less than the critical size, the probability of it breaking will be very low.

1.3.5. Limiting distribution (extended grinding)

Batch tests show that whatever the initial granulometry of the output is, we get, after a sufficiently long grinding period, the same granulometry limit. Furthermore, granulometry moves from being bimodal at the start to become subsequently unimodal (see Figure 1.1).

At the beginning, the density peak of fines is low. It then increases at the expense of the density peak of granules that gradually disappear.

Fines come from the conical area right below the compression surface of grains. Outside the cone, pieces are less fine and the residue gives coarse grains (granules). Fines build up with the number of compressions generating contact cones. There are more fines for extended grinding at low power than there are for brief grinding at high power.

1.3.6. Physics of grinding by collision (by percussion)

Let E_c be the kinetic energy of a griding body arriving on a particle bed. However, it can also be the kinetic energy of particles arriving on a target (anvil).

$$E_c = \frac{1}{2} m_m v^2$$

m_m: mass of the moving solid (kg)

v: speed at the time of collision (m.s^{-1}).

According to Becker *et al.* [BEC 01], kinetic energy is transformed into elastic energy that is itself distributed between:

– the moving solid

$$E_m = \frac{Y_f}{Y_m + Y_f} \times E_c$$

– the fixed solid

$$E_f = \frac{Y_m}{Y_m + Y_f} \times E_c$$

Y_m and Y_f: Young's modulus of a moving solid and a fixed solid

However, energy transferred to the fixed solid is partly transformed into heat in the ratio of Φ_f. It is the same for a moving solid with the coefficient Φ_m. Finally,

$\Phi = 0$ for a soft solid like limestone

$\Phi = 0.1$ for a hard rock like corundum

Finally, in Joules:

$$E_m = \frac{1}{2} m\, v^2 \left(1 - \Phi_m\right) \left[\frac{Y_f}{Y_m + Y_f} \right]$$

$$E_f = \frac{1}{2} m\, v^2 \left(1 - \Phi_f\right) \left[\frac{Y_m}{Y_m + Y_f} \right]$$

After the collision, the fragment dimension x_{50} (where underflow is 50%) depends on two parameters:

– energy E_f if the output is fixed (or energy E_m if the output is moving);

– specific energy W, a quotient of E_f or E_m by the mass of the product being treated.

For this, we refer to Figure 1.6 which resembles Figure 9 of Becker *et al.* [BEC 01].

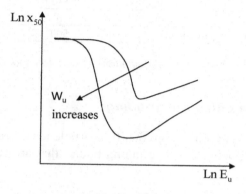

Figure 1.6. *Appearance of variations in x_{50} compared to E_u and W_u*

It is likely, but this has not been proven, that dimensions x_{10}, x_{20}, ... x_{90}, x_{100} represent variations of the same order.

1.3.7. *Falling speed in mills*

Air resistance does not act here on grinding bodies falling as these bodies are dense. If h is the fall height, the final speed is given by:

$$v_c = \sqrt{2g\,h}$$

g: acceleration due to gravity (9.81 m.s^{-2}).

Table 1.2 gives an approximate value for the final speed at collision:

– for either a grinding body on particles being fragmented;

– or for particles on a target.

Device	Speed (m.s^{-1})
Jaw or cone crushers	0.1–5
Ball or rod mills	0.5
Track mill	0.5–5
Roller mill	2.5
Hammer mill	50–100
Autogenous mill	8
Fluid energy mill	500

Table 1.4. *Order of magnitude for collision speeds*

1.3.8. *Physics of pinch roll grinding*

We suggest, for simplification, that the particle to be crushed is spherical and that it is stuck between two grinding bodies that are equal, spherical or cylindrical.

For the particle not to escape: the friction coefficient μ between the particle and the grinding body is such that:

$$\mu \ge tg\frac{\theta}{2}$$

Let us look at two planes tangent to two grinding bodies at two points of contact between the particle and each grinding body. The angle of these two planes is θ. It can be shown that

$$\cos\frac{\theta}{2} = \frac{D+e}{D+d_p}$$

D: diameter of the grinding body (m)

d_p: diameter of the particle (m)

e: minimum separation between grinding bodies (m)

Generally, μ is between 0.3 and 0.7.

During prolonged grinding, we find that selection speed for the output approaches a limit and this is much faster if [AUS 81a]:

– the solid concentration is high;

– the size of the output achieved is low.

Hence, we obtain a selection speed of zero for

d_p	% solid
10μm	65
1 μm	54

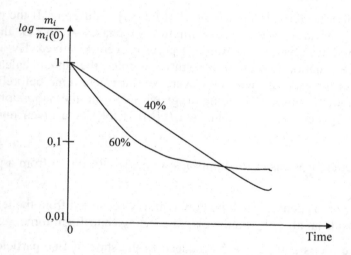

Figure 1.7. *Onset of the interval m_i by prolonged grinding*

For concentrations greater than 54%, the selection speed is not canceled out but is compensated by the lumping of many particles, so that the fineness of the grind reaches a limit. Therefore, we cannot expect to obtain any fineness with just any equipment.

Opoczky and Farnady [OPO 84] studied the way that selection speed tends to zero.

1.3.9. *Release of an ore: selectivity of a grinding procedure*

When pieces or grains of an ore contain inclusions of premium material, the grain size must be reduced to the size of inclusions in order to obtain particles that each corresponds to a single phase.

According to King [KIN 79], the level to which a phase is released, for a diameter of a given particle, is the fraction between these particles that contain more than a particular proportion of ore.

According to Barbery [BAR 92], the level to which a phase is released for a given diameter of a particle, is the fraction (by volume) of the particle corresponding to this phase.

The point of view of Peterson *et al.* [PET 85] is different. If the particles undergoing grinding are a binary mixture of phases A and B, they can, during grinding, give rise to Apure, Bpure or even ABmixed. By applying the balance equation to each of these three samples, the authors calculate the total output for each of them. However, we must determine beforehand the many parameters when grinding a given mass in the laboratory. This model was generalized by Mehta *et al.* [MEH 89] for a given number of components.

In practice, it is easier to simply *use selectivity* resulting from a property of the ore:

– If the ore is dense, it will be preferentially separated from the less dense gangue inside an air or hydraulic classifier using centrifugal force.

– If the ore is soft, it will be reduced to the state of fine particles much faster. As such, galena (hardness: 2.5 Mhos) will get to 200 μm while a gangue rich in quartz (hardness: 7 Mhos) will get to 600 μm. Naturally, during this procedure, galena and the gangue are combined in the ore that is being ground.

1.4. Grinding: a unit operation

1.4.1. *Parameters and fundamental grinding equations [AUS 71a]*

First, determine the size (diameter) of the constituent particles in a bulk solid. The classic method is to push the given bulk solid through a series of sieves. The sieves are overlaid and their openings (screen size, which is always square) decrease from top to bottom. These sieves are numbered from 1 to n starting from the sieve on top. Note that *this number system changes in the opposite direction to particle diameter*. Particle movement is achieved by vibrating the set of sieves.

In Figure 1.8, between sieves $i-1$ and i, the retained solid is said to belong to the interval i. The sieve n is not a sieve per se but a solid plate. The solid in interval n is said to be supersievable as it can contain particles that are as fine as we want.

Now, let us look at the grinding equation involving the mass m_i of particles in the interval i. For this, let us write that the change in speed of the mass m_i as the result of:

– Breaking the fraction k_i of the mass m_i during a unit of time. This fraction therefore disappears from the mass m_i.

– The inflow of fragments with size d_i from the breaking of particles j that are much bigger than the size d_i. We see that the erosion of particles i could be maintained in the interval i of these particles if the size of detached fragments is much smaller than d_i. In this case, $j = i$. Therefore, $b_{ii} \neq 0$.

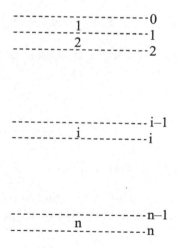

Figure 1.8. *Normal provision of a set of vibrating sieves*

The material balance equation for the interval i is written as:

$$\frac{dm_i}{dt} = -k_i m_i + \sum_{j=1}^{i} k_j b_{ij} m_j \qquad [1.1]$$

The coefficient b_{ij} represents the proportion of fragments coming from j where the size is d_i. We can write:

$$\sum_{i=j}^{n} b_{ij} = 1 \qquad [1.2]$$

In fact, the sum of fragments coming from j is equal to $k_j m_j$.

A very useful expression for b_{ij} is:

$$b_{ij} = \varnothing \left(\frac{x_i}{x_j} \right)^{\gamma} + (1-\varnothing) \left(\frac{x_i}{x_j} \right)^{\beta}$$

It involves an equation with three parameters and, in general:

$$0.2 < \varnothing < 0.5 \qquad 1 < \beta < 5 \qquad 0.5 < \gamma < 1$$

Such an expression corresponds to a standardized relationship. If it is not the case, we can try to make $\varnothing$ depend on x_j

$$\varnothing_j = \varnothing_o \left(\frac{x_j}{x_o} \right)^{\delta}$$

This means adding an extra parameter δ.

These analytical forms are clearly interesting as they provide a solution to the grinding problem when determining five or six parameters (for example, with a gradient method).

The combined overflow settled on top of the sieve's mesh i is, by definition:

$$R_i = \sum_{j=1}^{i} m_j$$

Noting from this definition:

$$m_i = R_i - R_{i-1}$$

The combined underflow totals everything that has passed through the sieve i:

$$P_i = \sum_{j=i+1}^{n} m_j$$

It follows that:

$$P_i + R_i = 1$$

1.4.2. *Coefficient for grinding speed: selection speed*

Let us consider a mill with a mass m_i of bulk solid made up of particles all with the same diameter d_i. When grinding begins, particles disappear because they were broken. By analogy of a chemical reaction of nth order, Patat and Mempel [PAT 65] write that the change in mass m_i is:

$$\frac{dm_i}{dt} = -k_i m_i^n \qquad [1.3]$$

In reality, it appears difficult, if not impossible, to establish an explicit grinding theory if we admit that n is not 1. In fact, integrations should be done numerically. However, fortunately for the vast majority of situations, the linearity hypothesis (n = 1) is verified.

Austin and his colleagues, as well as others, have written the speed equation:

$$\frac{dm_i}{dt} = -S_i m_i$$

S_i sets the proportion of the mass m_i of particles that disappear through breaking per unit of time. The coefficient S_i "selects" those particles that will be broken. Therefore, S_i is the *selection speed* for particles with a diameter d_i.

Integrating the speed equation:

$$Ln \frac{m_i}{m_i^{(o)}} = -S_i t \qquad\qquad [1.4]$$

If the linearity hypothesis is verified, then ln m_i is represented by a straight line as a function of time (line (a) in Figure 1.9).

NOTE.–

Austin proposes that:

S_i be called the relative selection speed

$m_i^{(o)} S_i$ be called the absolute selection speed

1.4.3. *Determining selection and breaking matrices*

Here, we will make a bibliographical list, in chronological order, of the methods that enable the extraction, from experience, of the value of parameters S_i and b_{ij}:

– Klimpel and Austin [KLI 70]

– Austin and Luckie [AUS 72b]

– Austin and Bhatia [AUS 72a]

– Luckie and Austin [LUC 72]

– Austin and Luckie [AUS 72c]

– Austin *et al.* [AUS 73b]

– Austin and Bhatia [AUS 73a]

– Klimpel and Austin [KLI 77]

– Austin and Bagga [AUS 81a]

– Gupta *et al.* [GUP 81]

– Klimpel and Austin [KLI 84]

NOTE.–

Austin, Shoji *et al.* [AUS 74] propose a new expression for the overflow versus time when milling a load R (x, t).

1.4.4. *Deviations from the linearity law [AUS 71]*

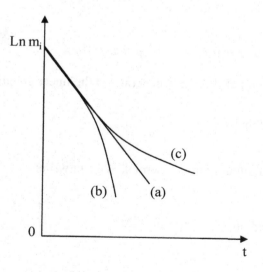

Figure 1.9. *Deviations from linearity*

The decrease over time of the fraction m_i during milling can be written as:

– A slowing down (curve c) with an accumulation of harder fractions (which break after other ones) or even by a "cushion effect" when there are a lot of fines that cushion the impact from a body undergoing grinding by letting the particle that was hit retreat within a bed of fines.

– An acceleration (curve b) because the largest particles were broken first, removing their purpose of being a protective shield for the mass of average-sized particles.

The selection speed S_i changes according to the particle size and increases according to the size. In the interval m_i, the largest particles break first and the others break much slower, which corresponds to the curve (c). However, we see that as the sample we are dealing with approaches a unique diameter d_i, it tends more toward a type (a) linear curve.

If the experiment provides type (b) or (c) curves, then Patat and Mempel [PAT 65, see 1.4.5] kinetics is valid. However, many theoretical and practical works have shown that only the linearity hypothesis accounts for a logical interpretation of grinding.

1.4.5. *Abandoning the linear hypothesis*

Patat and Mempel [PAT 65] generalized the linear equation by writing:

$$\frac{dm_i}{dt} = -S_i m_i^{q_i} + \sum_{j=1}^{i-1} b_{ij} S_j m_j^{q_j}$$

A numerical integration is essential. We can write:

$$\Delta m_i = \left(-S_i \overline{m}_i^{q_i} + \sum_{j=1}^{i-1} b_{ij} S_j \overline{m}_j^{q_j} \right) \Delta t$$

With:

$$\bar{m}_{ik} = \frac{1}{2}\left(m_{i,\,k-1} + m_{i,\,k+1}\right) \quad \text{and} \quad \bar{m}_{jk} = \frac{1}{2}\left(m_{j,\,k-1} + m_{j,\,k+1}\right)$$

1.4.6. *Zero-order grinding*

This only deals with the finest fines. Let us assume that they do not break, and the change in their total mass only comes from the fragmentation of the largest particles. As such:

$$\frac{dm_i}{dt} = \sum_{j=1}^{i-1} S_i b_{ij} m_j$$

1.4.7. *The distribution (or breaking) matrix*

The distribution matrix has as elements the coefficients b_{ij} that represent the proportion of fragments of size i coming from particles j.

For elements of this matrix:

$$b_{ij} = 0 \qquad \text{if} \qquad j > i$$

In fact, a fragment could have a size greater than that of the grain from which it comes. The matrix B is therefore a lower triangular matrix. These elements b_{ij} exist only for:

$$j \leq i$$

Finally:

$$\sum_{i=j}^{n} b_{ij} = 1$$

Let us define B_{ij} as the combined fraction coming from interval j and that, during sieving, falls under the interval $i - 1$. In other words:

$$B_{ij} = \sum_{k=i}^{n} b_{kj}$$

Naturally:

$$b_{ij} = B_{ij} - B_{i+1, j}$$

1.5. Distribution of residence times (continuous grinding)

1.5.1. *Parameters*

The distribution of residence times is defined using the following three parameters:

1) the average residence time τ that is hence defined as:

$$\tau = \frac{M}{F}$$

M is the retention, that is the constant mass of powder or pulp continuously present in the mill. It can be determined by weighing.

F is the flow of mass. This flow must be known.

2) the average μ of the distribution is given by:

$$\mu = \frac{\int_0^\infty t\, c(L, t)\, dt}{\int_0^\infty c(L, t)\, dt}$$

c (L, t) is the concentration of the marker coming out of the mill of length L at time t.

3) Variance s^2 of the distribution

$$s^2 = \frac{\int_0^\infty (t - \mu)^2\, c(L, t)\, dt}{\int_0^\infty c(L, t)\, dt}$$

NOTE.–

Hogg *et al.* [HOG 75] propose a relationship between the filling of the mill and the flow F that goes through it. However, this relationship is not predictable.

1.5.2. *Changes to parameters in the distribution of residence time*

1) Holdup M:

According to Gupta *et al.* [GUP 81] and, for wet grinding, holdup:

– is independent of the solid's bulk density;

– increases linearly with the flow of the pulp that is fed in and with the solid's load fraction in the pulp.

In all cases for wet grinding, holdup increases with [SWA 81]:

– the flow of solid (linearly);

– ball diameter;

– load of balls after resistance to the increase in pulp.

With balls, holdup is less compared with a rod mill.

Retention goes to a minimum in terms of rotation speed.

2) Average residence time τ

This decreases with the flow of solid and this decrease is greater when the pulp is more concentrated and when Bond's energy index is much higher.

3) Péclet's value uL/D and the dispersivity D [SWA 81]

Above 25% of the bed of grinding bodies in the mill's volume, Péclet's value no longer depends on the volume of the bed of grinding bodies.

Dispersivity does not depend on the size, nor on the material of the particles being processed [AUS 83b].

Dispersivity decreases with flow because the powder's motions are restricted due to higher holdup.

Dispersivity is much smaller for a rod mill than for a ball mill as the rod's axial motion is very restricted.

Retention is at a maximum when the rotation speed changes. In fact,

– at a low speed, there is a decrease in collisions;

– at a high speed, the cataract also decreases collisions.

1.5.3. *Austin, Luckie, Ateya [AUS 71] solution*

The equation connecting dispersion (D is dispersivity) and transport to speed u is written as:

$$\frac{\partial c}{\partial t} + u\frac{\partial c}{\partial x} - D\frac{\partial^2 c}{\partial x^2} = 0 \qquad [1.5]$$

The total quantity injected at the mill's input is:

$$Q = \int_0^\infty c(L,\ t)\,dt$$

A solution for equation [1.5] is, at a distance x from the input:

$$f(x,\ t) = \frac{c(x,\ t)}{Q} = \frac{u}{2\sqrt{\pi Dt}}\exp\left[-\frac{(x-ut)^2}{4Dt}\right]$$

and, at the output, since $u = L/\tau$ and $x = L$

$$f(L,\ t) = \frac{c(L,\ t)}{Q} = \frac{L}{2\tau(\pi Dt)^{1/2}}\exp\left[-\frac{L^2\left(1-\dfrac{t}{\tau}\right)^2}{4Dt}\right]$$

Average residence time is, by definition:

$$\tau = \frac{L}{u} = \int_0^\infty t f(t)\,dt$$

L: length of the mill.

But this solution does not work for limit conditions, that is at the input and output areas of the mill. These conditions are:

$$\frac{\partial c}{\partial x}\bigg|_{x=0} = 0 \quad \text{and} \quad \frac{\partial c}{\partial x}\bigg|_{x=L} = 0$$

The authors were therefore required to perform numerical integration. They give curves that help to determine dispersivity D by knowing the concentration profile at the output versus time when a short impulse is injected at the input.

NOTE.–

We wrote that $\tau = \dfrac{L}{u}$, but we can also write that $\tau = \dfrac{M}{F}$

M: actual mass in the mill (kg)

F: machine feed flow (kg.s^{-1})

1.5.4. Analogy with n identical mixers in a series

The mixer i's mater balance is written as:

$$V_u \frac{dc_i}{dt} = Qc_{i-1} - Qc_i$$

Where:

$$\tau_u = \frac{V_u}{Q} \quad \text{and} \quad \theta = \frac{tQ}{V_u} = \frac{t}{\tau_u}$$

V_u: volume of a mixer (m)

Q: flow moving through mixers (m^3.s^{-1})

τ_u: residence time in a mixer (s)

The balance becomes:

$$\frac{dc_i}{d\theta} = c_{i-1} - c_i$$

Let us call $\overline{c}(s)$ the Laplace transform of c. The derivation comes back to a multiplication by s

$$s\overline{c}_i = \overline{c}_{i-1} - \overline{c}_i \quad \text{or} \quad \overline{c}_i = \frac{\overline{c}_{i-1}}{s+1} \quad \text{and} \quad \overline{c}_1 = \frac{\overline{c}_0}{s+1}$$

From where:

$$\overline{c}_n = \frac{\overline{c}_0}{(s+1)^n}$$

The inverse of the Laplace transform is the problem's solution. It is given in equation [32.31] from Spiegel [SPI 92]:

$$F(\theta)d\theta = \frac{\theta^{n-1}e^{-\theta}}{(n-1)!}d\theta \quad \text{or} \quad f(t)dt = \frac{(1/\tau_u)^n t^{n-1}e^{-t/\tau_u}}{(n-1)!}dt$$

So, for a ball mill processing clinker, Austin, Luckie and Wightman [AUS 75] use:

$$n = 10$$

In more general terms, Levenspiel [LEV 62] gives:

$$n = \frac{\tau^2}{s^2} \quad \text{with} \quad \tau = n\tau_u$$

s^2: variance of the distribution law

τ: average residence time (s)

A laboratory mill is equivalent to a unique mixer (n = 1)

In this case:

$$f(t) = \frac{1}{\tau}e^{-t/\tau}$$

1.6. Solving mill and grinding circuit equations

1.6.1. *Transforming grinding equations by Reid [REI 65]*

After recalling Reid's results, we complete them by a practical procedure to calculate the coefficients a_{qi}.

According to Reid's equation [10]:

$$m_i(t) = \sum_{q=1}^{i} a_{qi} e^{-k_q t}$$

[1.6]

If we assume that the coefficients b_{ii} of the breaking matrix are not zero, we can write, by replacing (i-1) with i in the sum of Reid's equation [13]:

1) for $q \neq i$

$$a_{qi} = \frac{k_i b_{ii} a_{qi}}{k_i - k_q} + \sum_{j=q}^{i-1} \frac{k_j b_{ij} a_{qj}}{k_i - k_q}$$

Let us set:

$$\alpha q_i = 1 - \frac{k_i b_{ii} a_{qi}}{k_i - k_q}$$

That is:

for $q \neq i$ $\qquad a_{qi} = \frac{1}{\alpha_{qi}} \sum_{j=q}^{i-1} \frac{k_j b_{ij} a_{qi}}{k_i - k_q}$

[1.7]

Reid's result is equivalent to proving that α_{qi} is equal to 1.

2) for $q = i$

$$a_{ii} = m_i(0) - \sum_{q=1}^{i-1} a_{qi} \quad \text{with} \quad a_{11} = m_i(0)$$

[1.8]

This is Reid's equation 14 in [REI 65].

Let us now calculate, in a practical way, the coefficients a_{qi}.

According to recurring relationships 1.7 and 1.8, coefficients a_{qi} only exist for:

$$1 \leq q \leq i-1$$

Changing notations and substituting k_i for S_i that are the classical notations for selection speed for breaking.

We will distinguish several intervals where we will refer to j differently.

$q \leq l \leq i-1$

$$a_{qi} = \frac{1}{S_i - S_q} \sum_{l=q}^{i-1} S_l b_{il} a_{ql}$$

$q \leq m \leq l-1, i-2$

$$a_{ql} = \frac{1}{S_l - S_q} \sum_{m=q}^{l-1} S_m b_{im} a_{qm}$$

$q \leq n \leq m-1, j-2, i-3$

$$a_{qm} = \frac{1}{S_m - S_q} \sum_{n=q}^{m-1} S_n b_{in} a_{qn}$$

$\vdots \qquad \vdots \qquad \vdots$

$\vdots \qquad \vdots \qquad \vdots$

$\vdots \qquad \vdots \qquad \vdots$

$q = s = r-1 \quad$ and $\quad r-1 = 1$

$$a_{qr} = \frac{1}{S_r - S_q} \sum_{s=q}^{r-1} S_s b_{is} a_{qs}$$

By varying a from 1 to $i - 1$, we get all the coefficients of $a_{i\ell}$. The coefficient a_{ii} is, in and of itself:

$$a_{ii} = m_i(0) - \sum_{j=1}^{i-1} a_{ji}$$

Finally, we get equation [1.8].

Some authors involve a matrix T which occurs equally as its inverse. The usefulness of this procedure is not apparent.

When two relative speeds of selection S_j and S_q are very close, the previous method is dangerous and matrix integration is preferred (see section 1.6.5).

1.6.2. *Grinding a mass M and continuous grinding*

Equation [1.6] immediately gives the mass of the interval versus time

$$m_i(t) = \sum_{q=1}^{i} a_{qi} e^{-S_q t} \qquad\qquad [1.9]$$

With:

$$\sum_{i=1}^{n} m_i(t) = cste = M$$

If a continuous mill functions as a plug flow, then it only needs to use the transit time of the solid being processed in the mill as the milling time.

In reality, the residence time of particles in the machine obeys a distribution law that we can call $\varphi(t)$ frequency. This law is the same for all particles, or rather for all grinding processes leading to a product characterized by size intervals of $m_{ik}(t_k)$.

$$m_i = \sum_{k=1}^{\infty} \varphi(t_k) m_{ik}(t_k) t$$

Therefore:

$$m_i = \int_0^{\infty} \varphi(t) m_i(t) dt$$

However, according to Reid's calculation, we can write, for milling a load during time t, by replacing $m_{pi}(t)$ with $m_i(t)$ of the equation [1.9]

$$m_i = \int_0^{\infty} \varphi(t) \sum_{l=1}^{i} a_{qi} e^{-S_q t} dt = \sum_{l=1}^{i} a_{qi} I_q$$

With:

$$I_q = \int_0^{\infty} \varphi(t) e^{-S_q t} dt$$

1.6.3. *Calculating integrals I_s for continuous grinding*

We get two situations:

1) Austin, Luckie and Ateya [AUS 71b] give:

$$\varphi(t) = \frac{L}{2\tau(\pi Dt)^{1/2}} \exp\left[-\frac{L^2 \left(1 - \dfrac{t}{\tau}\right)^2}{4Dt} \right]$$

Gardner and Sukanjnajtee [GAR 73] give, from the expression for $\varphi(t)$ by Austin *et al.* [AUS 71b]:

$$I_q = \frac{L}{\tau\left[\dfrac{L^2}{\tau^2}+4DS_q\right]^{1/2}} \exp - \left[\frac{\dfrac{L^2}{\tau}-L\left(\dfrac{L^2}{\tau^2}+4DS_q\right)^{1/2}}{2D}\right]$$

D: dispersion coefficient (dispersivity) $(m^2.s^{-1})$

L: length of the mill (m)

τ: average residence time (in the mill)

$$\tau = \frac{M}{F}$$

M: mass solid in the mill (kg)

F: flow solid being processed $(kg.s^{-1})$

We find orders of magnitude for D in the publications of Swaroop *et al.* [SWA 81], Austin, Luckie and Ateya [AUS 71b] and finally in Austin *et al.* [AUS 83b].

2) More generally, it is suitable to liken a continuous mill to a series of perfectly agitated capacities m (see section 1.5.4)

$$\varphi(t) = \frac{\tau_u^{-m} t^{m-1} \exp\left(-\dfrac{t}{\tau u}\right)}{(m-1)!}$$

The residence time τ_u is the unit residence time in each capacity

$$\tau_u = \frac{\tau}{m}$$

Gardner and Sukanjnajtee [GAR 73] give the value I_q to the integral for this second expression for $\varphi(t)$

$$I_q = \frac{\tau_u^{-m}}{\left(\tau_u^{-1} + S_q\right)^m}$$

1.6.4. *Taking axial dispersion into account*

In such equipment, the granulometry of the output changes all along the axis. For the granulometric layer i, an assessment allows us to write (see equation 1.5 of section 1.5.3):

$$D\frac{d^2 m_i}{dx^2} - u\frac{dm_i}{dx} - S_i m_i + \sum_{j=1}^{i-1} b_{ij} S_j m_j = 0$$

D: axial dispersion coefficient (m²/s)

u: product axial speed

L: length of equipment (m)

τ: average product residence time (s)

$$\tau = \frac{M}{F}$$

M: mass of solid inside the equipment (holdup) (kg)

F: feed flow rate (kg/s)

Let us set

$$\eta = \frac{x}{L} \qquad P\acute{e} = \frac{uL}{D} \qquad \sigma_i = \frac{LS_i}{u}$$

We then obtain:

$$\frac{1}{P\acute{e}}\frac{d^2m_i}{d\eta^2} - \frac{dm_i}{d\eta} - \sigma_i m_i + \sum_{j=1}^{i-1} b_{ij}\sigma_j m_j = 0$$

This equation is a second-order differential equation that is homogeneous by 1 degree with respect to m_i and m_j and width constant coefficients. There are n such equations since i can vary from 1 to n.

For each of the equations, if there is not a grid classifier at the output, conditions at the boundaries are the same as for continuous liquid–liquid extractors:

For $\eta = 0$:

$$u\,f_i = u\,m_i - D\frac{dm_i}{dx}$$

Or even, under a semi-dimensionless form:

$$f_i = m_i - \frac{1}{P\acute{e}}\frac{dm_i}{d\eta}$$

For $\eta = 1$:

$$\frac{dm_i}{d\eta} = 0$$

If there is a gate, we can write:

$$um_i = (1-g_i)\left(um_i - D\frac{dm_i}{dx} \right)$$

That is:

$$g_i um_i = -D\frac{dm_i}{dx}$$

g_i: fraction of feed retained by the grid.

NOTE.–

Mika [MIK 76] suggests a matrix solution for the continuous milling problem. This solution is even applicable to vibrating mills. This author continues straight from the progression equation of the solid by introducing the axial dispersion coefficient into it.

1.6.5. *Matrix integration*

The material balance of the size interval i is written as:

$$\frac{dm_i}{dt} = -S_i m_i + \sum_{j=1}^{i-1} b_{ij} S_j m_j \qquad [1.10]$$

m_i: mass fraction of particles with a diameter between d_i and $d_i+\Delta d_i$. This fraction is simply called "fraction with diameter d_i"

m_j: fraction with diameter d_j

S_j: relative selection speed for the diameter d_j (s^{-1})

b_{ij}: fraction of mass m_j that is transformed into particles with diameter d_i. b_{ij} are distribution coefficients.

The total mass of fragments derived from the breakage of the fraction m_j is $m_j S_j$ with:

$$\sum_{i=j+1}^{n} b_{ij} = 1$$

n: number of sieves in the sieving device (see section 1.4.1).

n the differential equations [1.10] could be represented in a matrix form:

$$\frac{dM(t)}{dt} = -S\,M(t) + B\,S\,M(t) = (B-I)\,S\,M(t) \qquad [1.11]$$

M: granulometric vector of the mill's internal product

S: selection speed diagonal matrix. Elements s_i are constant.

B: distribution matrix of elements b_{ij}.

I: diagonal matrix unit.

The equation [1.11] easily integrates:

$$M(t) = \left[\exp(B - I)St\right]M(0) \qquad [1.12]$$

Remember that the matrix grouping exponent is:

$$\exp[(\mathbf{B} - \mathbf{I})\mathbf{S}t] = \sum_{k=0}^{\infty} \frac{1}{k!}[(\mathbf{B} - \mathbf{I})\mathbf{S}t]^k$$

The product of two matrices is very easy to program.

The transfer matrix D, is defined by:

$$M(t) = D.M(0) \qquad [1.13]$$

From the equations [1.12] and [1.13], this transfer matrix D is written as:

$$D = \exp\left[(B - I)St\right]$$

NOTE.–

Berkeley's team (see [GRA 69]) exposes the results synthetically. Therefore, it is beneficial to read this publication.

1.6.6. *Standardization hypothesis*

We assume that *the value of elements b_{ij} depends only on the ratio x_i/x_j.* In other words, when the diameter x_j of the mother particle changes, the sizes x_i of daughter particles remain proportional to the mother particle size.

We know that particle sizes are numbered in geometric progression because $\rho < 1$ from the interval 1 up to the interval n.

$$\frac{x_{i-1}}{x_i} = \frac{1}{\rho} = cste \qquad \frac{x_1}{x_2} = \frac{1}{\rho} \qquad \frac{x_2}{x_3} = \frac{1}{\rho} \qquad \frac{x_{n-1}}{x_n} = \frac{1}{\rho}$$

Therefore:

$$x_i = x_1 \rho^{i-1} \qquad x_j = x_1 \rho^{j-1} \qquad \frac{x_i}{x_j} = \rho^{i-j}$$

As a result, elements b_{ij} are a unique function of the difference $i - j$. Elements diagonal to the matrix B are therefore equal to the inside of each diagonal and the matrix B is written as:

$$= B = \begin{bmatrix} b_{11} & & & & & \\ b_{21} & b_{22} & & & & \\ \vdots & & \ddots & \ddots & & \\ \vdots & & & \ddots & \ddots & \\ \vdots & & & & \ddots & \ddots \\ b_{n1} & \cdots & \cdots & \cdots & b_{n,n-1} & b_{nn} \end{bmatrix} = \begin{bmatrix} b_1 & & & & & \\ b_2 & b_1 & & & & \\ \vdots & & \ddots & \ddots & & \\ \vdots & & & \ddots & \ddots & \\ \vdots & & & & \ddots & \ddots \\ b_n & \cdots & \cdots & \cdots & b_2 & b_1 \end{bmatrix}$$

Indices of the matrix on the right are equal to $i - j + 1$

Therefore, only one parameter is enough to identify the elements of the standardized matrix. We can then talk about mill function where the direct form is made up of b_k.

1.6.7. *Inversion of a lower triangular standardized matrix*

$$B = \begin{bmatrix} b_1 & 0 & 0 & \cdots & \cdots & 0 \\ b_2 & b_1 & 0 & \cdots & \cdots & 0 \\ b_3 & b_2 & b_1 & \cdots & \cdots & 0 \\ \vdots & \vdots & \vdots & & & \vdots \\ \vdots & \vdots & \vdots & & & \vdots \\ b_n & b_{n-1} & b_{n-2} & \cdots & \cdots & b_1 \end{bmatrix}$$

Let us call C the matrix B^{-1} the inverse of B. We must have:

B C = I

The inverse of a lower triangular matrix is also a lower triangular matrix.

$$C = \begin{bmatrix} c_1 & 0 & 0 & \cdots & \cdots & 0 \\ c_2 & c_1 & 0 & \cdots & \cdots & 0 \\ c_3 & c_2 & c_1 & \cdots & \cdots & 0 \\ \vdots & \vdots & \vdots & & & \vdots \\ \vdots & \vdots & \vdots & & & \vdots \\ c_n & c_{n-1} & c_{n-2} & \cdots & \cdots & c_1 \end{bmatrix} \quad \text{and } I = \begin{bmatrix} 1 & & & & \\ 0 & 1 & & & \\ 0 & 0 & 1 & & \\ \vdots & \vdots & \vdots & & \\ \vdots & \vdots & \vdots & & \\ 0 & 0 & 0 & & 1 \end{bmatrix}$$

The first column I is obtained by making the product of lines from B and the first column from C.

$$b_1 c_1 = 1$$
$$b_2 c_1 + b_1 c_2 = 0$$
$$\vdots$$
$$\vdots$$
$$\vdots$$
$$b_n c_1 + \cdots \cdots \cdots b_1 c_n = 0$$

We immediately deduct elements from C.

NOTE.–

Rogers [ROG 83] lists elements from the transfer matrix D:

$$p_i = \Sigma d_{ij} f_j$$

For the following mills:

ball mill for a load

continuous ball mill

air swept ball mill

track mill

Two rod mill.

1.6.8. *Recirculation processes*

For "open" circulation; a single pass through the mill could lead to "over-milling" for feed particles that have already satisfied the fineness specification. In addition, energy used for over-milling is wasted.

A classifier at the mill output will eliminate particles that *comply* with the specification (where the diameter is less than d_s) and will send particles that are too big (granules) back into the mill. With this process, the granulometry of the product is constricted. Figure 1.10 shows the corresponding circulation diagram.

If feed M of the system contains a lot of particles already at specification, it will be mixed with the output P coming out of the mill and processed in the classifier that will select particles Q with the correct size and return oversize granules for re-milling. We call this circuit "reverse recirculation" (see Figure 1.11).

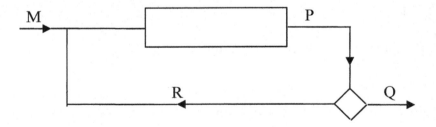

Figure 1.10. *Process with recirculation*

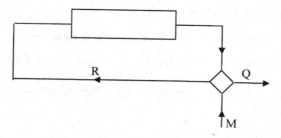

Figure 1.11. *Reverse recirculation*

Recirculated granules R are made up of the "circulating load" and the recirculation rate is the ratio:

$$\frac{\text{recirculated flow R}}{\text{feed flow M}} = \frac{\text{recirculated flow R}}{\text{production flow Q}}$$

The recirculation rate could, depending on the situation, vary from 0.5 up to 10.

We take M = Q = 1 in calculations. As such, the recirculation rate is equal to R.

Generally, the closed circuit (with classifier) works for machines with fine output. The open circuit (without classifier) could, as we know, lead to an energy wastage or an over-grinding of fines.

Grains with the greatest resistance accumulate in the mill cycle.

Kelleher [KEL 59] details the advantages and inconveniences associated with a closed circuit device.

1.6.9. *Classifier principles*

A classifier could be external or internal.

External classifiers are: screens, cyclones, hydrocyclones.

There is also an internal classification system in the mill. It is simply the use of an air current to pull the fines through the machine.

An ideal classifier separates:

– Particles with a diameter $d > d_s$. These are thus granules;

– Particles such that the diameter $d > d_s$. These are the fines.

The diameter d_s is the specification diameter (or the separation diameter or, finally, the cut-point diameter).

1.6.10. *Matrix calculation of a circuit in continuous operation [CAL 67]*

The circuit that we will study is given in Figure 1.12.

In what follows, we will specify certain points in the calculation of Callcott [CAL 67] as the latter was not always made very explicit.

We will use the following column vectors:

m, r, f, p, q

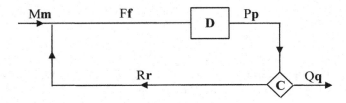

Figure 1.12. *Diagram of a given direct circuit*

D is the transfer matrix of the mill and C is the diagonal matrix of the classifier. In this matrix, the term c_i represents the fraction of mill output found in the interval i that is recycled through the mill's feed. At the limit, $c_i = 1$ for granules and $c_i = 0$ for fines.

$$Rr = PC\,p \qquad\qquad [1.14]$$

$$Qq = P(I\text{-}C)\,p \qquad\qquad [1.15]$$

$$Pp = FD\,f \qquad\qquad [1.16]$$

$$M=Q \tag{1.17}$$

$$Pp=Qq+Rr \tag{1.18}$$

$$Ff=Rr+Mm \tag{1.19}$$

Equations [1.15], [1.18] and [1.19] come from the law of mixtures.

From equation [1.17], we can define *a rate of recirculation*.

$$\rho = \frac{R}{Q} = \frac{R}{M}$$

If we work with Q and M equal to 1, then $\rho = R$.

From equations [1.14] and [1.19]:

$$Ff=PC\,p+Mm$$

Or even, from [1.15]:

$$F(I-C\,D)\,f=Mm$$

which can be written as:

$$f=\frac{1}{F}(I-C\,D)^{-1}Mm \tag{1.20}$$

But:

$$F=R+M$$

From where:

$$f=\frac{1}{R+M}(I-C\,D)^{-1}Mm \tag{1.21}$$

At the output of the circuit, from [1.9], [1.10] and [1.11]:

$$Qq=(R+M)(I\text{-}C)D\,f \qquad\qquad\qquad [1.22]$$

From [1.21] and [1.22]:

$$Qq=(I\text{-}C)D(I-C\,D)^{-1}Mm$$

But, from [1.17]:

$$q=(I\text{-}C)D(I\text{-}CD)^{-1}m$$

As such, the transfer matrix X *of the circuit* is defined by:

$$X=(I\text{-}C)D(I\text{-}C\,D)^{-1}$$

NOTE.–

Reverse Circuit

When the input Mm is made up of a large proportion of fines, it is wise to receive this input directly into the classifier that sends fines directly to the output.

Previous calculations remain valid as long as m is replaced with **C m** and to add **(I − C) m** at the output.

1.6.11. *Recirculation rate*

The equation [1.19] is written, by taking equation [1.20] into account:

$$Rr=Ff-Mm=\left[(I-C\,D)^{-1}-I\right]Mm$$

But:

$$\sum_{i=1}^{n} r_i =1 \quad \text{and} \quad \rho=\frac{R}{M}$$

The recirculation rate is therefore:

$$\rho = \sum_{i=1}^{n} \left[\left[(I - C\,D)^{-1} - I \right] m \right]_i$$

1.6.12. *Benefit of a circuit with a classifier*

With a classifier, a circuit conserves energy and sometimes uses a shorter mill.

Viswanathan and Mani [VIS 84] showed, using a simple example, that if the selectivity of the classifier relative to the granules increases, the recirculation rate decreases.

NOTE.–

We will find examples of circuit calculations in the publication by Callcott [CAL 67]. The work of Austin and Luckie [AUS 72d] exhaustively studies all possible circuit configurations. Particularly in case 4 of the latter authors, we find the simulation of Bond's procedure finds the energy index (see section 2.2). Finally, Austin, Luckie *et al.* [AUS 84] solved the problem with a mill using an air current passing through it, which entrains fines.

1.6.13. *Treating workshop's atmosphere*

Equipment must operate under a slightly lowered pressure to avoid ejecting dust outside.

A bag filter extracts the dust out of the air drawn into the mill.

A safety filter frame supporting a filter cloth treats the air before it is vented to the outside.

Sometimes, continuous control of the oxygen content in the mill's atmosphere prevents explosions.

A humidity control could prove to be important if the treated product is hygroscopic. Hygroscopicity aggravates the quality of the output.

Grinding Energy

2.1. Power and yields

2.1.1. *Order of magnitude of energies at play in grinding*

Industrially speaking, we measure the energy used in kilowatt hour per ton of ore. This energy increases very quickly with the desired fineness of the output exiting the machine.

Table 2.1 gives a very approximate idea of this energy expressed in kilowatt hour per ton (specific energy). Harris [HAR 66] provides an analogous table expressed in imperial units.

Grinding quality	Crushing	Average	Fine	Ultra-fine
Desired fineness	1 cm	1 mm	100 µm	< 10 µm
Specific energy	3–5	4.5–6	20–30	100–1,000

Table 2.1. *Specific energy to be consumed*

From the definition for grinding energy (see section 2.2.2), we realize that the energy to consume increases rapidly with the fineness of the output obtained. This is what is confirmed in Table 2.1.

For output sizes less than 0.1 µm, grinding will not be suitable, regardless of the method used. As such, to obtain particles with a size that is of the order of 10 nm, we proceed with obstructed crystallization, which means that we adsorb a constituent on the germ surface (normally organic) that serves as a barrier to new molecules attaching to the crystal's surface. Naturally, said barrier must subsequently be removed, that is from the adsorbent on the nanocrystal's surface. For this, it will be appropriate to use a solvent that is different from the mother liquor.

2.1.2. *Net power*

Power called the comminution unit's net power is the difference between:

1) shaft power when the unit is a processing product;

2) shaft power when the unit is working without a product.

Net power is the sum of both terms:

1) the power necessary to propagate cracks, that is useful power;

2) the power lost that corresponds to stresses visible in the product that are not followed by cracks; this term breaks down as heat.

Net power is not a strictly unique function of the nature of the product being processed and the rate of reduction, as it depends on the comminution mode, meaning ultimately the type of machine.

2.1.3. *Relationships between yields*

We can write:

industrial yield = fragmentation yield × mechanical yield

As long as we define:

$$\text{fragmentation yield} = \frac{\text{useful power}}{\text{net power}}$$

$$\text{mechanical yield} = \frac{\text{net power}}{\text{power consumed by the motor}}$$

2.1.4. *Useful power*

To characterize the fractionation of a solid, we measure the surface area of the solid before and after grinding. The difference ΔS in these surfaces characterizes the comminution operation.

The work $\mathcal{T}$ used in this test is provided by a mass M falling from a height h on a solid undergoing grinding.

There is no reason for ΔS to be proportional to work $\mathcal{T}$. In this test, we just assume that there is only one relationship (previously unknown) between these two magnitudes.

To determine useful energy in a comminution unit, we just measure the number (n) of falls of a mass M required to obtain the same value of ΔS as an industrial unit. Under these conditions:

nMgh = useful energy

Charles [CHA 56] examined the breakage of a given mass of solid undergoing a collision created by a mass falling. However, we typically do not measure the useful energy corresponding to a given machine.

The useful power of a machine is the product of useful energy and the flow of product mass.

NOTE.–

It is absurd to try to evaluate useful energy by the relationship:

$\gamma \, \Delta S$

γ: surface energy ($J.m^{-2}$)

ΔS: increase in the bulk solid's surface (m^2).

Basically, in this case, the yield of fragmentation is only a few percentage values.

2.1.5. *Industrial yield, industrial energy*

Useful power is the power required to propagate a crack. Industrial recovery η of a comminution unit is defined by:

$$\eta = \frac{\text{useful power}}{\text{shaft power}}$$

Unit	Yield $\eta\%$ (estimated)
Jaw or cone crushers	30
Hammer crusher	20
Roller mill	60
Ball or rod mills	20
Track mill	20
Autogenous mill	15
Hammer mill	10
Erosion mill	5

Table 2.2. *Industrial recoveries*

In reality, we do not typically measure useful power. This is why industrial recovery can only be estimated.

Table 2.3 gives the approximate industrial energy (consumed by the motor in a given time period) according to the size of particles produced for certain common materials.

$\log_{10} E$ is defined by $\pm 20\%$

$\log_{10} d$ is defined by $\pm 30\%$

Naturally:

$$E = 10^{\log_{10} E}$$

Material	$\log_{10} d$ (d in μm)	$\log_{10} E$ (E in kWh.tonne^{-1})
Glass	0.4	1.3
Inorganic pigments	0.2	2.7
Cocoa, milk powder	0.5	1.7
Sugar	0.5	0.3
Clinker	1.2	1.7
Flour grains	1.3–1.5	1.3–1.5
Iron ore	1.5	1.7
Pyrite	1.5	1.4
Plastic materials	2.2	2.7
Hard coal	2.3	0.8
Autogenous milling	3.4	1.2
Soft coal	3.6	0.7

Table 2.3. *Industrial specific energy*

2.1.6. *Grinding productivity*

1) Surface area of a solid particle:

The surface area of a solid sphere is:

$$\sigma_{sp} = \frac{\pi d^2}{\rho_s \pi d^3 / 6} = \frac{6}{\rho_s d_s}$$

More generally, particles are not spherical. The non-sphericity coefficient must therefore be applied which is given by the following ratio:

$$\varnothing_{NS} = \frac{\text{particle surface}}{\text{sphere surface of the same volume}}$$

Particle shape	$\varnothing_{NS}$
Sphere	1
Spheroid	1.05–1.15
Common salt (eroded)	1.20
Cube	1.235
Circular cut cylinder	1.37
Square cut cylinder	1.47
Thick disc (1 by 4)	1.45
Glass or sharp sand	1.52
Thick sheet (1 by 4 by 4)	1.56
Mica chip	3.57
Sharp anthracite	1.59

Table 2.4. *Non-sphericity coefficients*

Grinding an ore always produces sharp particles, except for spheroid particles released from comminution and that are not the outcome of a breakage.

The surface area of a solid particle is therefore:

$$\sigma_{pa} = \frac{6\varnothing_{NS}}{\rho_s d_p}$$

2) Surface area of a bulk solid:

Let us assume that this solid is distributed in n classes where each has a diameter d_{pi} (i from 1 to n), a density and a non-sphericity coefficient $\varnothing_{NSi}$.

Let us assume, in addition, that each class represents the mass fraction m_i in the bulk solid.

The surface area of the bulk solid $(m^2.kg^{-1})$ is:

$$\sigma = \sum_{i=1}^{n} \frac{6\varnothing_{NSi} m_i}{\rho_{si} d_{pi}}$$

3) Grinding productivity:

The net specific energy for grinding ($J.kg^{-1}$) is defined as:

$$\Pi = \frac{P}{W}$$

where

W: flow rate of mass in the processed ore ($kg.s^{-1}$)

P: net mill power ($J.s^{-1}$ = Watt)

Let us make F the bulk solid fed and P the output from the machine. The solid's surface created by the grinding procedure is in $m^2.s^{-1}$.

$$\left(\sigma_P - \sigma_F\right) W$$

Grinding productivity η_B is the ratio of the output's surface in a unit of time to the net power implemented P:

$$\eta_B = \frac{\left(\sigma_P - \sigma_F\right) W}{P} = \frac{\sigma_P - \sigma_F}{\Pi}$$

We notice that this definition relates to easily measurable quantities and is hence practical as opposed to notions of industrial recovery where inaccuracy is a known factor from the fact that it is impossible to correctly measure the energy that is actually dedicated to fragmentation.

Grinding productivity hence defined is measured in $m^2.J^{-1}$. A standardization of this notion and the way to obtain it will be most welcome.

2.1.7. *The mode of comminution's influence on energy consumed*

During a collision, fragments carry a significant amount of kinetic energy with them which is dispersed as heat in the surroundings. The energy recovery in collision mills will therefore be less than that in slow compression units.

If a layer of grains is slowly compressed, these will be subjected to highly variable types of stress and diverse intensities in a way that energy is lost as mutual friction and reversible elastic deformation that do not lead to fractionation.

Efficiency in grinding a layer of grains depends on the granulometry of the layer. For mono-dispersed granulometry, this efficiency will be mediocre. Inversely, if the grains are flooded with fines, it will equally be bad, as grains will therefore be subjected to quasi-hydrostatic compression. However, it will be excellent if the diameter of fines is ¼ of the grains that will undergo grinding. Finally, everything depends on the number of stresses and the intensity of the pressure load placed on them. Yet, the more that these loads are increased, the less they are intense.

By calculating Fairs' [FAI 53] results for limestone, barite and anhydrite, we find energy consumption (inverse of efficiency) for grinding in $J.m^{-2}$ according to the mechanism used, as presented in Table 2.5.

Collision (isolated mass or balls)	77
Hammer mills	104
Erosion	210

Table 2.5. *Energy consumption for grinding*

The collision speeds required depends on the desired fineness of output coming out of the collision machine, as shown in Table 2.6.

Fineness (μm)	Speed ($m.s^{-1}$)
100	60
10	100
1	300

Table 2.6. *Collision speeds and fineness*

2.1.8. *Motor power of machines*

Kelleher [KEL 59] gives, in Table 2 (p. 469) of his publication, a table that gives general usage and the electricity consumed by each comminution unit.

Power is expressed in hph.ton^{-1} (horsepower × hour.ton^{-1}):

$$\text{hph.ton}^{-1} = \frac{0.74572}{0.907} = 0.82 \text{ kWh.ton}^{-1}$$

For a machine turning at pilot scale, we can estimate that 18% of the power shaft is lost during transmissions. This percentage is lowered to 14% for a machine of industrial size.

2.2. Bond's energy index: ball or rod mills

2.2.1. *Background*

Let us look at the volume of a very big solid and let us reduce it to n particles of unit size d_p with a surface that is proportional to d_p^2 :

– according to Rittinger [RIT 67], the energy that is needed to be supplied to the solid is proportional to the total surface area of particles obtained:

being n d_p^2

– according to Kick [KIC 85], the energy supplied is proportional to the volume of particles:

being n d_p^3

Let us now report the energy of the true volume of the solid being processed, as n. d_p^3. Let E be this energy density:

– according to Rittinger $E = \dfrac{n\, d_p^2}{n\, d_p^3} = \dfrac{1}{d_p}$

– according to Kick $E = \dfrac{n\, d_p^3}{n\, d_p^3} = \text{cste}$

Kick's law, valid for particles greater than 0.5 cm, was used less than Rittinger's law, which proved to be interesting for particles smaller than 100 μm at the mouth of the grinder.

Bond [BON 52] made the hypothesis that the exponent of d_p is the arithmetic mean between Rittinger and Kick's values, being 0.5. In addition, it returns energy not to the volume but to the mass being processed and, for the specific energy, he writes:

$$W_d = \frac{E}{\rho_s} = \frac{K}{\sqrt{d_p}}$$

2.2.2. Definition of Bond's energy index

According to Bond and relative to grinding, a solid of infinite size has, per unit mass, a zero energy level.

However, Bond suggests in an empirical way that every particle of a finite size d has per unit mass a level of energy in the following form:

$$W_d = \frac{K}{\sqrt{d}} \qquad [2.1]$$

Finally, Bond introduces an energy level standard W_i that is the energy to be supplied to a unit mass of solid of infinite size in order to divide it into uniformly sized particles equal to 100 μm:

$$W_i = \frac{K}{100^{0.5}} \qquad [2.2]$$

Bond calls W_i the "work index", literally meaning "energy index". Naturally, W_i depends only, in principle, on the nature of the milled body. It is the reason why he provides W_i values (as well as for the true density) for most commonly used ores in Table IIIA on page 548 of [BON 61b].

Substituting K in both equations [2.1] and [2.2]:

$$W_d = W_i \sqrt{\frac{100}{d}}$$

More generally, to go from a bulk solid with particles of uniform size d_f to an output bulk solid with particles of uniform size d_p less than d_f, a grinding energy W_B must be supplied:

$$W_B = W_{dp} - W_{df} = W_i \left(\frac{10}{\sqrt{d_p}} - \frac{10}{\sqrt{d_f}} \right)$$

According to Bond [BON 54], the manufacturer, with their range of equipment, should offer the machine that is capable of supplying the power P that is capable of dealing with the flow of feed Q with:

$$P = QW_B$$

P: the machine's "useful" power (kW)

Q: mass of flow being processed (kg.s^{-1})

W_B: grinding energy (kg.s^{-1} of processed output)

NOTE (Units).–

In Table IIIA, Bond [BON 61b, p. 548] provides the standard specific energies expressed in kilowatt hour which he calls a "short ton" (whose value is 2,000 pounds).

$$\frac{kWh}{short\ ton} = \frac{3\ 600\ kilojoule}{0.90718\ metric\ ton} = 3.968.10^3\ kilojoule.ton^{-1}$$

$$1\ kWh\,(short\ ton)^{-1} = 3.968\ kJ.kg^{-1}$$

2.2.3. *Interpretation of laboratory tests (concept)*

The concept of measuring W_i is the following. We put a total mass M_T of ore to be milled in a small laboratory grinder. After a determined number of rotations N_T, we empty the grinder and sort the mass M_T on a sieve with an opening P_1. The mass of the underflow is $M_P = M_T - M_R$, where M_R is the mass of overflow. We complete the mass M_R with "fresh" ore to obtain M_T that we reload into the grinder.

We repeat the operation until the sieving of M_T gives a constant result. We calculate the average of the last three operations. The grinding energy used in one cycle is, in Joules:

$$E = P\tau$$

P: electric power consumed (measured in Wattmeter) (W)

τ: cycle duration (s).

Assuming that the mass M_T placed in the grinder has, at the start of each of its last cycles, 80% underflow with a size of d_f microns and that this same mass has, at the end of each of these cycles, 80% underflow with a size of d_p microns, Bond's standard specific energy W_i will be:

$$W_i = \frac{E}{M_T \left(\dfrac{10}{\sqrt{d_p}} - \dfrac{10}{\sqrt{d_f}} \right)}$$

In [BON 61a, pp. 381–382], Bond provides, a more complete description of the procedure for his measurements.

2.2.4. *Measuring the index energy*

This requires a special laboratory grinder called "Bond's jar".

The output to be processed is first reduced so that its grains are all of a size that is less than 3,360 μm with 80% by mass that is less than 2 mm.

Then, 700 g of this output, compacted following a standard procedure, are dry milled in the vessel (ϕ = 305 mm and L = 305 mm). The rotation speed of the vessel is 70 rev.mn^{-1}, indicating 85% of the critical speed. The balls load weighs 20.125 kg and is made up of a specific number of balls with sizes ranging from 12.7 to 38 mm.

The method consists of grinding the load for a short period of time (100–300 revolutions). We then sieve the load with a screen size x_T set in advance (that is, 300 μm for example) and replace the downflow with an equal mass of fresh feed. We repeat the operation until there is a constant ratio equal to 2.5 between the overflow and the underflow and the mass M_F of underflow obtained for the content of the mill is equal to a constant.

The length of time required for each operation must be determined by trial and error and the number of operations may, depending on the situation, vary from 7 to 15. The method is therefore not simple.

The energy index is therefore:

$$E_I = \frac{44.5}{x_T^{0.23} \times G^{0.82} \left[\sqrt{\dfrac{100}{x_P}} - \sqrt{\dfrac{100}{x_F}} \right]} \qquad (\text{kWh}/907\text{ kg})$$

x_T: sieve opening (μm)

x_F: size of 80% of the underflow for the product to be processed (μm)

x_P: size of 80% of the underflow for the output obtained (μm)

G: grams of underflow obtained on the sieve for each revolution of the jar; G is Bond's "grindability".

EXAMPLE.–

$x_T = 300$ μm $x_P = 100$ μm

$G = 1.5$ g $x_F = 1,000$ μm

$$E_I = \frac{1.1 \times 44.5}{(300)^{0.23} \times (1.5)^{0.82} (1 - 0.31)}$$

$$E_I = 13.7 \text{ kWh} / (907 \text{ kg})$$

NOTE.–

We can express Bond's index in kilowatt hour per metric ton by writing:

$$E_I^* = 1.1 E_I \qquad\qquad (\text{kWh}/(1000\text{ kg}))$$

2.2.5. *Bond's correction coefficients to apply to the energy index*

1) Rotation speed:

The rotation speed is expressed as a fraction of the critical speed, such that the centrifugal force balances the weight:

$$\omega_c = \sqrt{\frac{g}{r}} = \sqrt{\frac{2g}{D}}$$

The critical rotation speed in revolutions per minute is:

$$N_c = \frac{60}{2\pi}\sqrt{\frac{2g}{D}}$$

or

$$N_c = \frac{42.3}{\sqrt{D}}$$

D: diameter of the rotational mill's horizontal axis (m)

N_c: critical speed (rev.mn^{-1})

We characterize the rotation speed of a ball or a rod mill as the fraction φ of the critical speed N_c:

$$N = \varphi N_c$$

Bond's Table II [BON 54] gives a correction coefficient for the mill's shaft energy according to the latter's type and φ.

2) Milled output with a size that is less than 70 μm:

The reference value W_i must be multiplied by an enhancing coefficient [BON 54, p. 384].

3) Wet grinding and dry grinding:

Dry grinding requires more energy than wet grinding. For dry grinding, we must multiply the reference W_i by 1.3, which is obtained by wet grinding.

4) Reduction ratio R [BON 61a, BON 61b]:

Bond proposes an optimum reduction ratio for rod mills [BON 61b, p. 545]. If the real ratio deviates, then W_i must be multiplied by a correction coefficient (equation 26 in [BON 61b, p. 545]).

For a ball mill, if the reduction ratio becomes less than 3 (target grinding of concentrates), the energy index W_i must be multiplied by a given coefficient given by the author's equation 27 [BON 61b, p. 545].

If we want to reduce a very large compact solid ($d_f = \infty$) into particles with diameter $d_p = 100$ µm, the required specific energy becomes equal to Bond's energy index W_i that is measured in kilowatt hour per short ton (907.18 kg) of solid. Values for this energy index will be provided for many ores [BON 60].

Note that carbon is missing from Bond's list. This issue was dealt with by Chandler [CHA 65]. The applicable standard method for carbon is Hardgrove's [HAD 32] test that Chandler [CHA 65] describes.

NOTE I.–

In practice, it is very difficult to make use of a bulk solid when the size of all its particles is equal to d_p. This is why Bond defines the sizes d_f and d_p as screen openings that let through 80% of the feed's solid mass and grinder output, respectively.

NOTE II.–

Bond [BON 54] proposed correspondent relationships between his energy index and magnitudes characterizing the capacity during grinding given in other texts written by Bond himself [BON 49].

2.2.6. *The specific energy consumed in grinding*

According to the foregoing, we could proceed in the following way.

The specific energy is given by:

$$E = E_I^* \left[\sqrt{\frac{100}{x_P}} - \sqrt{\frac{100}{x_F}} \right] \prod_{i=1}^{7} C_i \qquad (\text{kWh/ton})$$

1) Grinding coefficient:

dry $C_1 = 1.3$

wet $C_1 = 1$

2) Open circuit coefficient:

This coefficient depends on the mass fraction of the output that is less than 75 μm.

%	C_2
50	1.035
60	1.05
70	1.10
80	1.20
90	1.40
92	1.46
95	1.57
98	1.70

Table 2.7. Correction coefficient

3) Size of mill coefficient:

$D \leq 3.81$ m $C_3 = (2.44/D)^{0.2}$

$D > 3.81$ m $C_3 = 0.914$

4) Oversize coefficient of the feed:

$$C_4 = 1 + \frac{\left[(E_1 - 7)(x_F - x_o) \right]}{x_o R}$$

R: the desired reduction ratio (x_F/x_P)

x_o: the "optimum" size

x_F and x_P: feed and output sizes (in μm) for 80% underflow.

– for ball mills:

$$x_o = 4000 \sqrt{\frac{13}{E_I}} \qquad (\mu m)$$

– for rod mills:

$$x_o = 16000 \sqrt{\frac{13}{E_I}} \qquad (\mu m)$$

5) Output fineness coefficient:

$$C_5 = \frac{x_P + 10.3}{1.145 x_P}$$

This coefficient must not be greater than 1 and is only applicable to:

$$x_P < 75 \mu m$$

6) Rod mill coefficient:

$$C_6 = 1 + \frac{(R - R_o)^2}{150}$$

where the optimum reduction ratio R_o is given by:

$$R_o = 8 + \frac{5L_B}{D}$$

L_B: rod length (m)

D: mill diameter (m)

7) Ball mill coefficient:

$$C_7 = \frac{2(R - 1.35) + 0.26}{2(R - 1.35)}$$

This coefficient is only applicable when R < 6.

EXAMPLE.– Let us consider by dry grinding 40 ton.h^{-1} of ore in a ball mill that enters at 5 mm and exits at 200 μm. The apparent density of the ore is 1.5 and its energy index E_i is equal to 15. We estimate that the diameter of the mill is less than 3.8 m.

$$C_1 = 1.3 \qquad C_2 = 1$$

$$C_3 = \left(\frac{2.44}{3}\right)^{0.2} = 0.96$$

$$x_o = 4000\sqrt{\frac{13}{15}} = 3724 \ \mu m$$

$$C_4 = 1 + \frac{(15-7)(5000-3724)}{3724 \times \left(\frac{5}{0}.2\right)} = 1.11$$

$$C_5 = 1 \qquad C_7 = 1$$

$$E = 15 \times 1.1 \left[\sqrt{\frac{100}{200}} - \sqrt{\frac{100}{5000}}\right] \times 1.3 \times 0.96 \times 1.11$$

$$E = 22.86 \times (0.71 - 0.14) = 13 \ kWh \ / \ ton$$

$$P_a = 40 \times 13 = 520 \ kW$$

2.2.7. Power at the ball mill shaft

$$P_a = 7.33 \ J \ \varnothing (1-0.937J)\left(1 - \frac{0.1}{2^{9-10\varnothing}}\right) d_B L \ D^{2.3} \qquad (kW)$$

d_B: true density of grinding bodies (dimensionless)

ϕ: fraction of the critical speed

J: fraction of the mill's volume occupied by grinding bodies and their interstitial vacuum (for spherical balls, the fraction of vacuum is equal to 0.4)

L: mill length (m).

This formula is Bond's empirical formula.

EXAMPLE.–

$D = 3$ m $L = 4.5$ m $J = 0.3$

$d_B = 7.8$ $\o = 0.75$

$$P_a = 7.33 \times 0.3 \times 0.75 \left(1 - 0.937 \times 0.3\right) \left(1 - \frac{0.1}{2^{9-7.5}}\right) \times 7.8 \times 4.5 \times 3^{2.3}$$

$P_a = 1.64 \times 0.72 \times 0.96 \times 439$

$P_a = 498$ kW

2.2.8. The influence that ball and rod mill dimensions have on net power

The major portion of energy is used to raise balls against the weight of a height with a magnitude equivalent to the diameter of the mill. This energy is proportional to $\rho D_{CB}^3 D$ for a single ball.

D_{CB}: diameter of grinding bodies (balls) (m)

D: mill diameter (m)

ρ_B: true density of balls (kg.m^{-3})

For a given value of the filling rate of the bed of balls in the mill, the number of balls is proportional to $D^2 L / D_{CB}^3$, where L is the length of the mill.

The mass that is increased per unit time is proportional to the rotation speed of the ferrule, indicating the output having a critical speed of $42/D^{0.5}$ with the fraction φ of this speed.

As such, the mechanical power at play P_m is proportional to:

$$P_m \sim \left(\frac{42\varphi}{D^{0.5}}\right)\left(\frac{D^2 L}{D_{CB}^3}\right)\left(\rho_B D_{CB}^3 D\right)$$

where:

$$P_m \sim \varphi \rho_B D^{2.5} L$$

Austin [AUS 73a, AUS 73b] gave the first simple justification.

We could use the same reasoning for a rod mill but the energy needed to raise a rod will be proportional to $\rho D_{CB}^2 L D$ and the number of rods will be $\rho D_{CB}^2 L D$ and the number of rods would be $D^2 L / D_{CB}^2 L$. We obtain:

$$P_m \sim \left(\frac{42\varphi}{D^{0.5}}\right)\left(\frac{D^2 L}{D_{CB}^2 L}\right)\left(\rho_B D_{CB}^2 L D\right)$$

$$m \sim \varphi \rho_B D^{2.5} L$$

The formula is the same for the rod mill and the ball mill, but that for ball mills is probably more reliable.

NOTE.–

The relationships assume that the reduced speed and filling parameters are constant.

3

Ball and Rod Mills

3.1. Introduction

3.1.1. *Ball mills*

The reduction ratio ranges from 20 to 200 depending on the situation. The size of the feed must not exceed 2–4 cm.

The body (cylindrical) of the mill (horizontal axis) is divided into two or more sections by perforated walls.

The inner wall of the body (the ferrule) is equipped with shield plates that are subject to erosion and can be replaced.

In cascade operation mode, pieces roll on an inclined plane formed by the output, while in cataract operation mode, pieces are projected into the air.

Loose balls occupy 20–50% of the ferrule's volume. The output must occupy the entire open space between balls. The fraction of the volume of space between balls is 0.4, if the latter are all of the same size and less than 0.4 in the opposite case.

3.1.2. *Operation principle*

The ball mill is a cylindrical drum (or cylindrical conical) turning around its horizontal axis. It is partially filled with grinding bodies: cast iron or steel balls, or even flint (silica) or porcelain bearings. Spaces between balls or bearings are occupied by the load to be milled.

Following drum rotation, balls or bearings rise by rolling along the cylindrical wall and descending again in a cascade or cataract from a certain height. The output is then milled between two grinding bodies.

Ball mills could operate dry or even process a water suspension (almost always for ores). Dry, it is fed through a chute or a screw through the unit's opening. In a wet path, a system of scoops that turn with the mill is used and it plunges into a stationary tank.

3.1.3. *Dry operation and wet operation*

Dry milling is not used very often. It is suitable when:

– subsequent operations require a dry output;

– a dry output is being prepared and drying after wet grinding will be costly.

Arguments in favor of dry milling are as follows:

– wear on grinding bodies is on average five times less with a dry operation than with a wet operation;

– when feed characteristics change, air classifiers are much easier to control than hydraulic classifiers.

The reasons why a wet path is often required are as follows:

– the yield of dry mills decreases very quickly when the output's moisture exceeds 1%. Wet output agglomerates, balls and granules are covered in a layer of adhesive and plastic fines that cushions and lowers the force exerted on the output. In addition, the product circulates poorly in the mill. For these reasons, a hot air scan is often performed which requires an efficient dust removal facility;

– wet, the concentration of solid pulp must be such that the pulp's viscosity reaches 0.2 Pa.s. A surfactant allows for a higher solid content without increasing the fracture limit of the pseudo-plastic pulp. Production is hence increased;

– with a wet path, grains circulate well and do not re-agglomerate. In addition, it seems that resistance to fragmentation decreases in water. The result is that the energy consumption is greater with dry milling than with wet milling. It is on average 30% greater;

– air classifiers consume more energy than hydraulic classifiers and they are less efficient;

– wet handling and classification operations are much easier;

– wet production is 1.45–2 times more than dry production.

3.1.4. State of particles in emerging output [GAU 26]

Soft ores such as mica, pyroxene, hornblende and, to a lesser degree, feldspar wear very quickly on their surfaces. Quartz wears 100 times slower.

Generally speaking:

– fines (d_p < 100 μm) wear less easily than granules. In fact, fines have, in each particle, less faults that generate cracks;

– the outcome from grinding fines is more angular and less rounded than particles coming from granules.

The cascading fall of ore particles is produced by mill rotation. This repeated falling blunts sharp edges and tips. But this falling effect quickly diminishes if the size of the particles in the feed decreases.

This effect of eroding edges and tips is proportional, for each particle, to the particle's surface. The detached fragments are all less than 10 μm.

The outcome of the erosion phenomenon is that the frequency distribution of particle sizes produced is bimodal, meaning that it shows two maxima:

– one for fines, which is quite spread out;

– another for granules, which is very narrow.

In reality, the important parameter is *the ratio ρ of the size of the feed to the size of balls.* Let ρ_c be the critical value of this ratio.

If $\rho < \rho_c$, the distribution is the same for equipment with jaws and rollers and rods. It is characteristic of cracking and fractures running through particles, meaning a fragmentation that involves all particle sizes. The granules obtained are polyhedrons and fines are shards with shapes very different from a compact spheroid.

If $\rho > \rho_c$, the distribution is bimodal. Granules are rounded and worn from mutual grain erosion.

In principle, the ratio ρ_c varies from 1/6 to 1/10.

When we dramatically increase the grinding time of a given load with $\rho > \rho_c$:

– maximum fines displace toward smaller sizes;

– rounding (blunting) of granules increases;

– the frequency of intermediary sizes decreases.

Inversely, if $\rho < \rho_c$, all grains are fragmented starting with granules and, finally, we get fines.

The value of ρ_c depends on the mill's rotation speed.

If a particle's pinch angle between two grinding spherical or cylindrical bodies is low, the particle will be broken (shattered). Inversely (much larger particle), the particle will escape ball action and will only be eroded.

When the grinding time of a mass increases and if $\rho < \rho_c$, we go from Figure 3.1(II) to Figure 3.1(I). In other words, the "cavity" (we can also say the "valley") increases in depth. The granulometry frequency becomes bimodal (with two maxima).

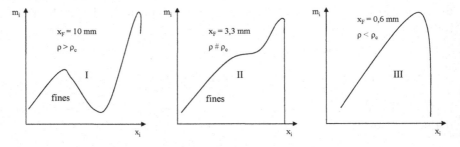

Figure 3.1. *Granulometry frequency of a ball mill*

For a continuous mill, plot III can only be obtained if the feed has a size that is less than the limit that depends on the size of balls, meaning $\rho < \rho_c$.

If this is not the case, balls only break the intermediary fraction and only blunt the sharp edges of granules (Figure 3.1(II)). We can relate this to the notion of the viewing angle (see roller mills).

Granules produced are rounded when the fed grain is too big (Figure 3.1(I)). They would have been worn by erosion.

3.2. Ball size

3.2.1. *Ball size calculation*

The shape of balls is a purely theoretical problem as they wear and take on a similar aspect to that of a sphere (or at least a pebble). In addition, it is found that only the spherical shape is the most selective, indicating that it gives a reduction rate that is higher for granules than for fines. In fact, this form corresponds to the maximum in the mass/surface ratio. Yet, granules are broken by balls falling (kinetic energy) and fines by the friction of balls against each other. This friction is as developed as the surface of grinding bodies is large. We note that, in a mill, balls move on their own by rotating

in the same direction as the rotation of the mill. This rotation becomes faster and faster as they get closer to the periphery.

This ball rotation creates a shearing force that favors fine grinding.

Selection of the balls' diameter making up the initial load must be such that the added amount breaks only the largest grains in the feed. If balls are too big, meaning they provide a low contact surface with the product, the amount of contact leading to fractionation will be reduced and productivity will be mediocre. Inversely, if balls are too small, contact without fractionation takes place between them and large pieces (granules).

For a given particle size, Figure 1.5 in Kelsall *et al.* [KEL 67/68] gives changes in S_i in terms of ball diameter. S_i is the relative selection speed for particle size x_i (see section 1.4.2). Austin *et al.* [AUS 76] have also studied this question.

In the case where balls do not have a uniform size, we admit that the selection speed S_i is a weighted average with weighted fractions of ball sizes:

$$S_i = \sum_k m_{Bk} S_{ik}$$

Ball dimension must be adapted to the granulometry of the feed. The curve below shows the phenomenon's appearance:

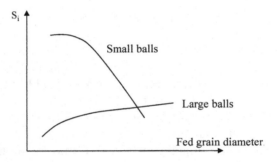

Figure 3.2. *Grain and ball size's influence on selection speed*

The Nordberg Manufacturing Company's formula lets us calculate a suitable diameter for steel balls or steel rods:

$$D_B = k \left(\frac{x_F E_I}{\varphi} \right)^{\frac{1}{2}} \left(\frac{\rho_s}{\sqrt{D}} \right)^{1/4}$$

With:

k = 0.0275 for balls

k = 0.02454 for rods

In this formula:

D_B: diameter of balls or rods (mm)

x_F: feed diameter with 80% underflow (µm)

E_I: Bond energy index (kWh. $(907$ kg$)^{-1}$, meaning kWh.ton^{-1})

φ: fraction of the critical speed

ρ_s: true density of ore (kg.m^{-3})

D: mill diameter (m)

EXAMPLE.–

We will use the Nordberg Manufacturing Company's formula

$$D = 0.61 \text{ m} \qquad \emptyset = 0.8 \qquad d_F = 3{,}000 \text{ µm}$$

$$\rho_s = 3{,}500 \text{ kg.m}^{-3} \qquad E_I = 8 \text{ kWh.}(907 \text{ kg})^{-1}$$

$$D_B = 0.0275 \left(\frac{3000 \times 8}{0.8} \right)^{\frac{1}{2}} \left(\frac{3500}{\sqrt{0.61}} \right)^{\frac{1}{4}}$$

$$D_B = 39 \text{ mm}$$

If the ball material is no longer steel (pebbles), the new diameter D_G conserves the unit mass:

$$D_G = D_B \left(7.8 / d_G \right)^{1/3}$$

d_G: density of the ball material.

Given that a fraction of space equals 0.4, the apparent density of balls is (in kg.m^{-3}) as given in Table 3.1.

Steel	4,500
Flint	1,600
Alumina	2,400

Table 3.1. *Apparent density of balls*

3.2.2. *Ball size according to Bond*

Bond [BON 58] gave an expression for the ball diameter. This expression is given in the following form (using Nordberg's notation):

$$D_B = 0.021 \left(\frac{d_f}{K} \right)^{1/2} \left(\frac{d\,W_i}{\varphi \sqrt{D}} \right)^{1/3}$$

d_f: diameter of feed screen for 80% underflow (μm)

d: density of output with respect to water

K: constant given in the author's Table II

D: mill internal diameter (m)

D_B: ball diameter (m)

φ: percentage of critical speed

3.3. Operation parameters

3.3.1. *Rotation speed*

The critical rotation speed is the speed in which the total load in the mill remains fixed to the wall because of the centrifugal force:

$$\omega_c^2 \frac{D}{2} = g, \quad \text{where:} \quad \omega_c = \sqrt{\frac{2g}{D}} \qquad \left(\text{rad.s}^{-1} \right)$$

Or even:

$$N_c = \frac{60}{2\pi} \sqrt{\frac{2g}{D}} = \frac{42.3}{\sqrt{D}} \qquad \left(\text{rev.mn}^{-1} \right)$$

D: mill diameter (m)

If the speed is 0.7 N_c, we obtain the cascade operation that is suitable for soft outputs. Balls rise and then fall back down by rolling on top of each other.

At a speed of 0.8 N_c, the operation is called cataract. Balls fall down again in free fall following a more or less parabolic trajectory. This rule applies to hard products.

In general, the solid surface created by fragmentation goes through a maximum for $N = 0.75 \ N_c$.

The relative speed φ is the ratio between the real speed and the critical speed. Very generally:

$$0.7 \leq \varphi \leq 0.75$$

3.3.2. Mill filling rate

The ratio of the shaded area to the mill's section (Figure 3.3) is:

$$J = \frac{1}{2\pi} \left[2\beta - \sin 2\ \beta \right]$$

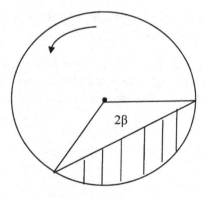

Figure 3.3. *Filling the mill*

Since balls are generally spherical, the porosity of the ball bed is 0.4 (for rods, it would be 0.2). The fraction of the volume 0.4 J held by the product is

U which is a little less than 1. Finally, the fraction f of the mill volume held by the product is:

$$f = 0.4 \text{ J} \times \text{U}$$

For example, if:

$$2\beta = 150° = 2.62 \text{ rad} \qquad \text{U} = 0.9$$

$$J = \frac{1}{2\pi}[2.62\text{-}0.50] = 0.34$$

$$f = 0.4 \times 0.34 \times 0.9 = 0.122$$

The optimum value of J is approximately 0.3–0.4.

NOTE.–

Gupta *et al.* [GUP 81a, GUP 81b] studied the impact that the solid pulp concentration has on the mass of the solid present in the mill.

3.3.3. *Pulp concentration and selection speed*

Tangsathitkulchai and Austin [TAN 89] studied the impact the pulp concentration on the relative selection speed.

In zone I of Figure 3.4, the viscosity (or rather the consistency) of the pulp increases and the ball bed is brought higher into the mill's rotation. The fall height of grinding bodies is increased, as a result, as much as the relative selection speed S_i.

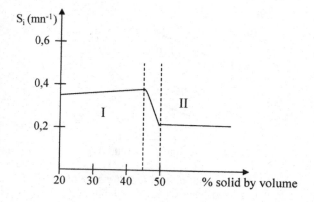

Figure 3.4. *Impact that the solid has on the relative selection speed*

For 45–50% solid, S_i drops suddenly because the pulp adheres to the wall where the breaking frequency is low.

In zone II, the increase in particle concentration is compensated by the increased consistency of the pulp and S_i remains practically a constant.

NOTE.–

Kawatra and Eisele [KAW 88] give the principle of the continuous measurement of the pulp's apparent viscosity.

3.3.4. *Pulp concentration and granulometry distribution*

Tangsathitkulchai and Austin [TAN 89] studied the impact the solid concentration on the granulometry distribution curve by plotting logP against log d where P is the combined underflow for particles with diameter d.

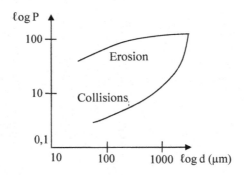

Figure 3.5. *Impact that the type of breaking has on the distribution curve*

Attrition produces a variety of fine particles, while breaking by collision produces more substantial fragments and fewer fines.

A higher concentration corresponds to a heavy pulp consistency that will form a thick layer on the wall that will absorb collisions by "wearing" particles but not breaking them. It is the characteristic of erosion. We could call this effect the "moisture cushion".

A lowered viscosity (low concentration) allows the ball to make a cascading fall in such a way that breaking occurs by *collision* between two balls to give much bigger fragments compared with erosion.

3.3.5. *Parameters for filling ball mills*

J is the fraction of the volume Ω of the mill held by the ball bed, including the spaces between balls.

If 0.4 is the porosity ε of the ball bed in the absence of an output to be milled, the available volume for this output is 0.4 JΩ.

U is the powder or pulp's usage rate for the available volume:

$$U = \frac{\text{powder or pulp volume based on mill volume}}{0.4\,J}$$

When the mill is overloaded, we find $U > 1$. In reality, balls separate from each other and we always get $U = 1$. However, in a fictitious and conventional way, we express the overload as $U > 1$, with ε always equal to 0.4. The orders of magnitude for the parameters U and J are:

$$0.5 < U < 1.5 \qquad\qquad 0.2 < J < 0.5$$

NOTE.–

The feed flow of a continuous mill has its significance:

– if it is high, filling the mill will be high with an output of fines and a loss of energy;

– if it is low, the output will be reduced by the same amount.

3.3.6. *Mill additives*

Additives could be classified as follows:

1) Liquid water or water in the vapor state attacks silica to form a more voluminous siloxane that opens the lips of a crack and activates grinding.

Some organic liquids or their vapors reduce the attractive forces between particles and promote grinding. Others reduce resistance to fracture (see Locher and Seebach [LOC 72]).

2) Some additives reduce flocculation, as the energy will be unnecessarily used to smash flock. On the other hand, particle dispersion maintains a distance between them through electrical repulsion, which reduces interparticle friction. As such, for wet milling:

– flocculation is hampered;

– dispersion is favored.

We refer the reader to Chapter 6 of [DUR 16] on soil destabilization.

3) Certain additives lower the viscosity of the pulp by the same mechanism as described previously by reducing interparticle friction. Low pulp viscosity prevents the centrifugal effect on the wall in ball mills. Fuerstenau and Venkataraman [FUE 85] studied this effect and developed a theory for balls fastening to the wall by demonstrating the existence of a critical viscosity with a magnitude of 5,000 centipoises. In fact, for a narrow interval of solid concentration in the pulp, this latter goes from a Newtonian state to a pseudoplastic state with a critical stress less than that for which there is no deformation.

4) Flotation additives that foster foaming are disastrous for grinding.

5) Some additives for dry milling form complexes on the solid's surface and hinder powder from coating balls.

Götte and Ziegler [GÖT 56] and especially Somasundaran and Lin [SOM 72] have deepened these considerations.

3.4. Flow in the mill

3.4.1. *Evacuation of pulp in wet milling*

– Overflow: A unit that operates as such is simple and not very sensitive to fluctuations in the feed system.

– Grid: We can search for a crude classification of output by placing a grid at the exit. Grid slots have a width that can range between 5 and 25 mm. With a grid, as the holdup of product in the mill is much lower, as a result, so is its residence time. The result is that the circulating load could be much higher compared with a runoff system. In addition, the grid allows for a ball charge that is as big as we want it to be. In fact, balls are not at risk of exiting the mill. The flow that a grid mill is capable of is 10–20% greater than the flow in a runoff mill. Energy consumption is equal in both. Yet, the grid system only has advantages. In fact, replacing worn grids is not free and it may clog the grid to require heavy cleaning.

3.4.2. *Flow of pulp through the grid*

The assumption is that the grid is pierced with holes of diameter δ and that the fraction of the surface held by holes is ε_g.

In Figure 3.6, the gray surface A is the section of the volume held by the ball-pulp combination.

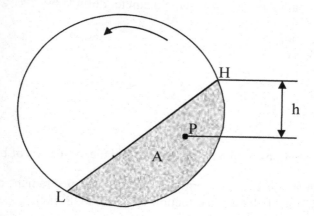

Figure 3.6. *Hydrostatic pressure*

The flow of pulp that runs through a hole located in P is:

$$q_u = 0.62 \frac{\pi}{4} \delta^2 \sqrt{2\,gh} \qquad\qquad \left(m^3 . s^{-1}\right)$$

h is the difference in dimension between point P and the highest point held by the pulp, which is point H.

If dA is an element of the gray area, the surface occupied by holes is $\varepsilon_g dA$ and the flow dQ running through the surface dA is:

$$dQ = 0.62\varepsilon_g dA \sqrt{2g\,h}$$

To find the total flow, just integrate the surface A:

$$Q = \int_A dQ$$

The angle that the line HL makes with the horizontal is about 35°.

3.4.3. *Height gradient of the pulp along the length of the mill*

We use the ball's average harmonic diameter defined by:

$$\frac{1}{\overline{D}_B} = \sum_i \frac{m_{Bi}}{D_{Bi}}$$

With:

$$\sum_i m_{Bi} = 1$$

m_{Bi}: mass fractions with ball diameter D_{Bi} in the total mass of balls.

We will then use Ergun's formula for the pressure gradient of the pulp passing through the ball bed (see section 1.1.2 of [DUR 16]).

$$\frac{dP}{dL} = 150 \frac{\mu V}{D_B^2} \frac{(1-\varepsilon)^2}{\varepsilon^3} + 1.75 \frac{\rho V^2}{D_B} \frac{(1-\varepsilon)}{\varepsilon^3}$$

ρ: pulp density (kg.m^{-3})

μ: pulp viscosity (Pa.s)

ε: fraction of free volume between balls (porosity)

V: pulp speed in an empty shaft (m.s^{-1})

P: pulp pressure (Pa)

L: mill length (m)

The gradient dh.dL^{-1} of the height of the pulp is such that:

$$\frac{dP}{dL} = \rho g \frac{dh}{dL}$$

By equating these two expressions for dP.dL^{-1}, we find the value for dh.dL^{-1}. We deduce the change in dimension Δh of the pulp between the mill's entrance and the exit, by considering the dimensions of the point H in Figure 3.6.

3.4.4. *Volume of powder in the mill in terms of flow*

This volume is characterized by the ratio f (see section 3.3.2).

Hogg *et al.* [HOG 75] suggests the following type of relationship:

$$f^{1/n} = fc(\theta)^{1/n} + \frac{K\varepsilon_B Fx}{\rho_a \cot g\alpha wR^4}$$

F: mass flow $(kg.s^{-1})$

α: dynamic talus angle (rad)

w: rotation speed $(rad.s^{-1})$

R: cylinder radius (m)

x: distance at exit (m)

ρ_a: apparent density of powder $(kg.m^{-3})$

K and n are two dimensionless empirical coefficients.

We see that the relationship between flow rate and retention is very different in dry milling and wet milling.

3.5. The selection matrix

3.5.1. *The selection function (diagonal matrix)*

It is a question of selecting the fraction S(x) of the proportion of size x particles that were broken in a unit of time.

Let us now consider a population of particles of various sizes located between two plane metallic surfaces. Let us assume that the two surfaces approach each other. The particles that will be the first to be subjected to a clamping pressure will be those whose *relative* height compared with the others will be high. As the surfaces approach each other, smaller and smaller particles will be crushed.

In other words, the probability of a particle to be crushed depends on its size x_i.

But the probability of being crushed will also be an increasing function of the useful power P_u of the mill.

In other words, the selection function $S(x)$ will be the product of a function $f(x/x^*)$ and a function $g(P_u)$ of the useful power P_u.

Finally, we obtain:

$$S(x) = f(x)g(P_u)$$

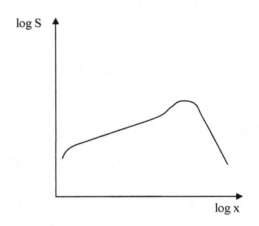

Figure 3.7. *Appearance of variations in the selection function for a given P_u (according to Weller et al. [WEL 88])*

3.5.2. *Selection speeds of ball mills*

According to Austin and Brame [AUS 83a], the selection function which is a diagonal matrix is written as (their equation 11):

$$1 \le i < n \quad S_i = f(x_i) \left(\frac{1}{1 + \left(\dfrac{x_i}{\mu}\right)^\Lambda} \right) \left(\frac{e^{-1.2U}}{1 + 6.6J^{2.3}} \right) \left(\frac{(\varnothing_c - 0.1)}{1 + \exp\left[15.7(\varnothing_c - 0.94)\right]} \right) \quad 1 \le i < n$$

$$\mu \propto \frac{d^2}{D^{N_2}} \quad \text{with} \quad 0.1 < N_2 < 0.2$$

In addition, the authors describe the precautions to be taken when:

– milling is dry or wet;

– blade readers are effective or not.

Trial tests are always required and authors describe how to extrapolate from a trial scale with a diameter D_T to an industrial mill with a diameter D.

3.5.3. *Extrapolation from trial to industrial scale*

Welles *et al.* [WEL 88] show that, from a trial, a simple translation of the log k_i = f(log x_i) curve in the (log k_i, log x_i) plane provides a good prediction of a new mill's performance. This is moreover what Austin and Brame's [AUS 83a] method suggests, but the method of the latter is credited to demonstrate the influence of each of the operation variables, which the method by Weller *et al.* [WEL 88] does not.

It is very important to note that the matrix for selection speeds depends on the mill's characteristics while the matrix for breaking depends only on the product being handled and, if the latter is homogeneous, the matrix for breaking could be standardized (see section 1.6.6).

3.5.4. *Effect of delay on breaking*

The selection speed could decrease during grinding for several reasons:

1) Fines coating balls. To verify this, they need to be cleaned periodically. Contact with coated balls results in insufficient stress to provoke fragmentation.

2) Non-homogeneous characteristics of the output. The softest grains fractionate first while the hardest grains are slow to fragment. This can be verified by setting aside the non-fragmented output after a set time. If the behavior of this reserve during grinding is the same as that of the overall feed, it means that the product being processed is homogeneous.

3) Agglomeration of fines. By incorporating radioactive fines to the feed, we can verify that they do not agglomerate.

4) The size limit corresponding to the product–mill pair is achieved. Small size grains then have plastic behavior and fragment with difficulty. If this is the case, we must note the difference in behavior during grinding by bringing the apparatus to the temperature of liquid nitrogen as small size grains then have elastic behavior.

5) Cushion effect. Excessive quantities of fines give grains a soft cushioning reaction in response to ball action. In fact, under a compression force, fines redistribute better between granules, which reduces the porosity and the output's volume without there being fragmentation. If we remove fines after a set time of non-continuous grinding, we see that the fragmentation speed goes back to its initial value. The cushion effect is produced if there is too much product in the mill.

To express the reduction in fragmentation speed $S_i (\tau)$ with time τ, we can try to set:

$$S_i(\tau) = S_i(o)K(\tau) \quad \text{with} \quad K(\tau) < 1$$

Interest in the coefficient $K(\tau)$ is that it is independent of index i, i characterizing the size of grains.

3.5.5. *Cushion effect in dry milling*

Look at the ultrafine particles, that is those particles with a diameter less than 10 μm. Here, we call these particles microscopic particles.

With such particles, powders obtained have a very high porosity. In fact, particles line up in more or less ramified chains and are more or less interconnected. The resultant effect resembles a steel frame where all the iron has been arranged haphazardly. We understand that the void fraction in volume for such a structure is high (0.7 or 0.8), while the void fraction for loose spherical balls is 0.4. The structure of powders is stable and does not collapse because the bond between the two particles is quite strong to mainly neutralize the weight of these particles. The force of the interparticle bond is due to van der Waals forces that we express quantitatively using Hamaker's law (see section 6.3.1 in [DUR 16]).

When a grinding body (ball) falls on a powder bed, the latter compresses by absorbing a significant amount of the ball's kinetic energy. This lost energy was dispersed in a volume that is much greater than the tiny hemisphere that included the stress in a plane in contact with a sphere. A hit particle could move back and disperse its kinetic energy in a powder bed.

When particles to be milled are made up of a mass greater than 15% of microscopic powder, the probability of fragmentation by collision or crushing could decrease dramatically.

Then, when a grinding body arrives on a powder bed, it destroys the powder's structure by compressing it like a sponge and, as a result, the interstitial air is expelled. But the permeability of a set of microscopic particles is very weak and air trying to leak into this environment brings particles from the structure with it, whence it is likely to release a cloud of dust.

3.6. Wearing of a mill's internal surfaces

3.6.1. Wear mechanisms

The mass of metal disappeared from wear during grinding could reach a thousandth of the mass of the processed solid. Two distinct phenomena compete in wearing:

1) Abrasion, (or erosion) which is a purely mechanical phenomenon, arises from the cutting action of hard particles of an ore following collision or compression. Surfaces hence eroded could demonstrate the following aspects:

– grooves;

– chips;

– scratches.

2) Corrosion only takes place in wet milling. This is why moisture wear is six to seven times greater than dry wear. A corroded surface could show:

– craters;

– punctures.

Corrosion could be fostered by:

– the existence of stress leading to cracks;

– the presence of oxygen that reacts electrochemically:

Cathode $\quad \dfrac{1}{2}O_2 + H_2O + 2e^- \quad \longrightarrow \quad 2OH^-$

Anode $\quad$ Fe $\quad\quad\quad\quad\quad \longrightarrow \quad Fe^{++} + 2e^-$

Overall $\quad \dfrac{1}{2}O_2 + H_2O + Fe \quad \longrightarrow \quad Fe(OH)_2$ precipitates

In the presence of oxygen, sulfides could be dangerous:

$$2O_2 + MS + 2H_2O \rightarrow M(OH)_2 + H_2SO_4$$

For example, we are aware of the devastating effects of dilute sulfuric acid.

3.6.2. *Magnitude of wear on internal surfaces*

A kilogram of worn shielding largely corresponds to about:

– 330 kWh in a dry way;

– 66 kWh in a wet way.

Wear is much higher in a wet operation from the fact that chemical reactions take place (especially when the product contains sulfides).

A kilogram of worn balls corresponds to:

– 75 kWh in a dry way;

– 15 kWh in a wet way.

Naturally, wear depends on the hardness of the processed output (see Appendix). We will consume 0.25 kg of steel for one ton of limestone and this is 10 times more for quartz.

3.6.3. *Remedies for wear*

The addition of chrome, manganese or molybdenum hardens the steel on grinding surfaces.

When metallic debris is banned, ceramics can be used (flint bearings, corundum cement and, perhaps, porcelain armor plates).

As for diaphragms, grids and especially classifiers (cyclones and hydrocyclones), we could use a polyurethane coating (easy to apply) and, sometimes, a rubber coating. In fact, the strong resilience that these materials have allows them to be resistant to collisions.

A neutral atmosphere (nitrogen) hinders oxygen's action.

A sufficiently concentrated pulp, hence viscous, covers balls and spreads over armor plates. As such, erosion is minimized. However, a viscous pulp impedes grinding.

3.7. Rod mills

3.7.1. *Description*

It is a ball mill with a cylindrical body where balls have been replaced with rods.

Rods have a 5–10 cm diameter and are shorter than the mill that is 15 cm to prevent blockage. We can use the formula from the Nordberg Manufacturing Company to calculate the rod diameter. In this case, the constant k has the value 0.02454 (see section 3.2.1).

The fraction of space between rods is 0.2 and the apparent rod volume takes up 35–40% of the mill's volume.

The rotation frequency is generally close to 70% of the critical value.

3.7.2. *Characteristics of rod mill output*

Fines are angular. Granules are equidimensional and are not tabular. Their shape is polyhedral.

While two adjacent rollers turn in the opposite direction in a roller mill, two adjacent rods turn in the same direction in a rod mill. Owing to this, transversal cracks (perpendicular to the direction of the applied crushing force) are rare.

There is an absence of transversal cracks made by rod mills on processed particles. In fact, these cracks create rather large fragments.

The speed of rod mills is slow compared with the speed of ball mills. The result is that rods roll and they do not collide with each other. The largest particles are pinched between two rods with a friction coefficient that is between 0.3 and 0.7. As such, fines are protected from the rod's action. The result is a *narrower* granulometry distribution for the output.

We see that the mill's deafening noise at the beginning of operation becomes more discreet over time when grinding is accomplished.

Granulometry of the output obtained is much narrower compared to a ball mill. Granules separate rods, fines travel through the space that is created without being over-milled and granules, which are subjected to a separation force, are fragmented. In addition, the free space between rods decreases as the output moves toward the exit.

The size of the feed is often between 1 and 4 cm. The underflow has 80% of the product and a size ranging between 0.5 and 5 mm. The most common reduction ratio has a magnitude of 25.

3.7.3. *Modeling*

Shoji and Austin [SHO 74] performed discontinuous tests with limestone, for which fragmentation function was not standardized. They noticed that, for each granulometry portion i, the fragmentation speed S_i had two limiting values:

– the value that they call standard, in the presence of a high proportion of grains with a size greater than x_i;

– a much greater value appears when the size i becomes the maximum size of the load.

In tests by Shoji and Austin [SHO 74], their conclusions confirm quantitatively that the preferred grinding of granules is with rod mills.

Hence, using analytical expressions for matrices S_i and B_i, we can assume that the models used for bar mills are equally applicable for rod mills.

3.7.4. Performance

Rod mills are used in an open circuit between crushing and the ball mill. They often operate in wet and where the pulp contains up to 50% solid by mass.

This machine is equally very applicable to dry where it could take a feed that has achieved 6% moisture. As such, when the product is sticky, rod mills are preferred over ball mills.

Rod mills have an industrial yield that is less than that of a ball mill, which explains the fact that balls have a much larger grinding surface than rods. The power needed to operate a rod mill could exceed 30% of the power used in a ball mill.

When they are worn, rods must not stretch but break in rather small pieces, so that they can exit the mill.

The energy consumed by rod mills could be from 10 to 30% higher than the energy consumed by bar mills and this is for comparable performance.

A rod mill is recommended if we want to properly grind large grains, because the ball mill will not attack them as well as rod mills will.

Crushers and Grinders
Except Ball Mills and Rod Mills

4.1. Crushers

4.1.1. *General points about crushers*

Primary crushers are characterized by the maximum size of boulders that they can accept and secondary crushers are defined by their reduction ratio.

The reduction ratio R is defined by:

$$R = \frac{\text{feed size}}{\text{product size at outlet}}$$

This ratio ranges from 4 to 8 and is moderate, which explains the low energy consumption per unit mass of product.

The presence of fines and, especially, moisture, could promote the formation of lumps, leading to a decrease in capacity and an increase in energy consumed.

Bond [BON 61a] gives an expression for jaw and gyratory crushers to obtain the size of the product obtained in millimeters. This size d_P corresponds to 80% underflow.

$$d_p = d_o \left(0.04 W_I + 0.40 \right)$$

d_o: opening at outlet (mm).

Bond [BON 61a] equally gives the size d_P for the cone crusher.

The fragmentation function obtained by crushing is expressed in the following form:

$$B_{i,j} = \varnothing \left[1 - e^{-\left(\frac{x_i}{x_j}\right)^u} \right] + (1 - \varnothing) \left[1 - e^{-\left(\frac{x_i}{x_a}\right)^v} \right]$$

It is an expression with four empirical parameters: u, v, Ø and x_a.

The first term recognizes pure and simple crushing and the second recognizes intergranular abrasion. The parameter x_a is the upper limit for fragment sizes obtained from abrasion.

This expression was used for cone crushers but we must not rule out the fact that it could equally be valuable for gyratory and jaw crushers and especially autogenous mills, where each one could replace one crusher followed by a mill.

4.1.2. Single effect jaw crusher

1) Principle

The machine has two jaws, opening toward the top for the feed. One jaw is fixed and the other jaw is mobile. The upper part of the mobile jaw has a circular movement created by an eccentric. The lower part of the mobile jaw has a recess at the bottom where an oscillating flap rests, whose function is to maintain a roughly constant distance between the two jaws. In fact, this distance defines the size of the crushed product and it is regulated by adjusting the flap.

When the mobile jaw moves forward, fragmentation is produced by pressure. This latter could be exerted directly: jaw on product or even product on product (self-grinding). When the jaw retreats, the product descends by gravity and the process repeats itself a dozen times up until the outlet. The proportion of pore spaces decreases as it descends. The distance between jaws at the outlet is the device's "adjustment" and corresponds to approximately 85% of the product's underflow.

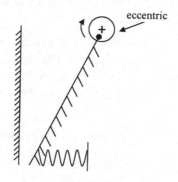

Figure 4.1. *Drive chain of a single-toggle crusher*

Two jaws in an open position form an angle that varies between 20 and 28° according to the situation. This angle is significantly smaller when the product is hard, moist and rounded and it has a weak friction coefficient. In fact, boulders should not escape the jaw's grip.

With all manufacturers, the L/G ratio of the width of the jaws at their opening on the feed side is equal to 1.25.

The frequency of oscillations could range between 50 and 750 cycles per minute depending on the substance's elasticity. Energy consumption is from 0.5 to 1.5 kWh.tonne^{-1}.

2) Interest and Performances

Advantages are as follows:

– a hydraulic device allows for adjustments when in operation;

– the jaw's wear plates could be turned around from top to bottom or even exchanged between themselves, which extends their lifespan;

– active surfaces could be adapted to the product. A smooth surface allows for a greater capacity than an embossed surface and it is preferred for soft products; an embossed surface eliminates flat grains; finally, a convex surface reduces the capacity but prevents jamming;

– the single-effect jaw crusher is suitable for products that are moderately abrasive (whose Mohs hardness does not exceed 5), as boulders grind each other (auto-grinding); however, we will see that the double-effect crusher is more suitable for very hard products;

– the reduction rate ranges from 4 to 7 depending on the adjustment. For a low rate, flow is obviously much higher. The optimum reduction rate is 5. The maximum "mouth" opening is 1.5 m, but it happens when the device is much bigger than the scale; the minimum size that we could hope for the product is approximately 4 cm;

– the device's capacity does not exceed 1,000 tonne.h^{-1}.

To finish, we note that the jaw crusher is not very suitable for moist and sticky products because there will be jamming. The solution involves using a hammer crusher.

4.1.3. Double effect jaw crusher

1) Principle

The upper part of the jaw pivots around a fixed point. The lower part has a forward and backward horizontal motion created by a system involving two connected flaps and one eccentric flap. When the flaps are extended one against the other, the opening at the base of the device is minimum. A single turn of the eccentric leads to two back-and-forth motions of the mobile jaw, where we get the name "double effect".

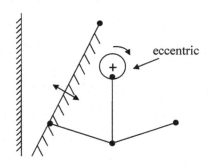

Figure 4.2. *Drive chain of a double-effect crusher*

2) Interest and Performances

Jaw motion in a single-effect crusher involves a vertical component that leads to significant friction wear. Inversely, the mobile jaw of a double-effect device has an essentially horizontal displacement and the friction is minimum. The double-effect principal is therefore used for very hard products such as ferrochromes and ferroalloys.

The double-effect crusher, which is clearly much more expensive than a single-toggle crusher, is reserved for limited use (very hard products).

4.1.4. *Characteristics of jaw crushers*

We will use the work of Rose and English [ROS 67].

The maximum size of boulders that the device can treat is given by:

$$x_F = 0.9\, G\ (m)$$

G is the maximum distance separating jaws in their upper part. It is the maximum dimension of the "mouth" of the machine.

The "landing" (or the run) of the mobile extremity of the moving jaw is given by:

$$J = 0.05\, G^{0.85}\ (J \text{ and } G \text{ in } m)$$

An upper limit of the reduction ratio is:

$$R = \frac{\text{mouth opening}}{\text{minimum opening at outlet}} = \frac{G}{S}$$

R is generally found to be between 4 and 7.

The practical frequency of the mobile jaw's strike is:

$$N_p = 280 \exp\left(-0.21\, G^3\right)\ (cycle.mn^{-1})$$

The critical frequency where, between two strikes, the product has just enough time to descend is:

$$N_c = \frac{47}{J^{\frac{1}{2}}} \left(\frac{R-1}{R} \right)^{\frac{1}{2}} \qquad\qquad \left(\text{cycle.mn}^{-1} \right)$$

4.1.5. *Jaw crusher flow rate [ROS 67]*

This flow rate is given, in tonnes per hour, by:

$$W = 5553\beta L J^{\frac{1}{2}} \left(S + \frac{J}{2} \right) \left(\frac{R}{R-1} \right)^{\frac{1}{2}} .x_F \left(1 + 1.9\ e^{-90\frac{J}{G}} \right) \varnothing \Psi$$

For granite and quartz: $\beta = 1$ (denser products)

For carbon and coke: $\beta = 0.55$ (lighter products)

L: equipment's width (m)

   $L = 1.25 \times G$

Ø depends on the granulometry of the feed through the intermediary of the parameter:

$$\alpha = \frac{\text{max. size} - \text{min. size}}{\text{average size}}$$

For $\alpha \leq 0.28$, $\varnothing = 0.4$

Otherwise, we will consult the table below:

A	Ø
0.28	0.40
0.40	0.65
0.66	0.84
1.20	0.88

Ψ depends on the strike frequency, from which it is understood that the flow is maximum at the critical frequency:

If $\quad N_p < N_c \quad \Psi = N_p/N_c$

If $\quad N_p > N_c \quad \Psi = N_c/N_p$

4.1.6. *Net power*

The net power is given by:

$$P_N = 1.35 \times I_E \times W \times \left(\sqrt{\frac{100}{x_P}} - \sqrt{\frac{100}{x_F}} \right) \ (kW)$$

x_F and x_P: feed and product sizes for 80% underflow (in μm).

I_E: Bond's energy index

EXAMPLE.–

$G = 1.5 \ m \quad R = 5 \quad \beta = 1$

$\alpha = 0.40 \quad I_E = 13 \ kWh.short \ ton^{-1}$ (1 short ton = 907 kg)

$x_F = 0.9 \times 1.5 = 1.35 \ m$

$J = 0.05 \times 1.5^{0.85} = 0.07 \ m$

$S = 1.5/5 = 0.3 \ m$

$N_p = 280 \times exp\left(-0.21 \times 1.5^3\right) = 138 \ mn^{-1}$

$$N_c = \frac{47}{(0.07)^{\frac{1}{2}}} \left(\frac{5-1}{5} \right)^{\frac{1}{2}} = 158 \ mn^{-1}$$

$L = 1.25 \times 1.5 = 1.87 \ m$

$$W = 5553 \times 1 \times 1.87 \times (0.07)^{\frac{1}{2}} \left(0.3 + \frac{0.07}{2}\right)\left(\frac{5}{5-1}\right)^{\frac{1}{2}}$$

$$\times 1.35 \left(1 + 10.9 e^{\frac{-90 \times 0.07}{1.5}}\right) \times 0.65 \times \frac{138}{158}$$

$$W = 900 \text{ t.h}^{-1}$$

$$P_N = 1.35 \times 13 \times 900 \times \left(\sqrt{\frac{100 \times 5}{1.35 \times 10^6}} - \sqrt{\frac{100}{1.35 \times 10^6}}\right)$$

$$P_N = 212 \text{ kW}$$

4.1.7. *Characteristics of jaw crushers and product at outlet*

We will find the following results in Rose and English's [ROS 67] publication:

– upper limit of the reduction ratio;

– flow rate that the equipment can manage;

– net power.

According to the authors, the maximum boulder size that the crusher can treat is given by:

$$x_F = 0.9 \text{ G}$$

G is the maximum distance separating jaws from their upper part. It is the opening of the machine's "mouth".

This crusher's action is similar to the action in roller mills. However, jaws give a product that is much richer in fines than rollers.

With smooth jaws or simply worn ones, the product descends in small successive columns and seems to possess, at the outlet, angular features.

4.1.8. *Description of a gyratory crusher*

The equipment's capacity has the shape of a conical trunk with the tip at the bottom and is called a bowl or a chamber. The chamber acts as the fixed jaw and it is covered in wear plates called "concaves".

The inner mobile surface of the chamber is made up of a conical trunk with its tip at the top. It is called a nut and it is also covered in armored plates. It follows from the above that the open space between grinding surfaces continues to retreat as the product descends, just like in a jaw crusher. The nut is suspended by the upper extremity of its axle to a ball bearing that rests on a crosspiece. The lower extremity of the nut's axle is moved by a circular movement of an eccentric system and this axle then sweeps the surface of a small angled cone.

The nut turns freely around its axle without any mechanical moving device, but only from this axle's rotation and friction on the product. As such, the nut rolls on the product without sliding. This movement regulates wear on active surfaces.

Through a complete rotation of the axle, the nut has made a revolution around the chamber and each of its points has come nearer and has moved away from the concaves in a movement that is very similar to the movement of jaws. The product descends by gravity and, each time it passes the nut, gets reduced into grains that are increasingly fine.

The angle between the chamber and nut surface is close to that of the jaws of a jaw crusher. Nevertheless, primary crushers have a smaller angle so that they do not let boulders escape, in the same way that the grip angle of roller crushers is also limited.

The size of the product leaving the base of the equipment is fixed by adjusting the minimum distance separating the nut and the chamber. This adjustment is made by changing the height of the nut in relation to the chamber. The smaller the adjustment, the finer the product, but flow rate is reduced. A lower radial run of the nut creates a granulometry that is more uniform and has less fines.

This crusher's mode of action is therefore comparable to the action of jaw crushers, but, here, the movement of grinding surfaces is continuous and the risk of blockage is minimum. A loading at "full mouth" is possible.

The rotation frequency of the axle of the nut changes according to the situation between 300 and 500 rev.mn^{-1}.

4.1.9. *Gyratory crusher performance*

– The gyratory crusher can, just like a jaw crusher, accept boulder sizes close to 1.5 m (for the largest equipment).

– This crusher can handle products with a Mohs hardness that is not necessarily less than 5.

– As is the case for jaw crushers, there is not any mutual sliding on grinding surfaces, but simply rolling of the nut on concaves. Wear is then uniform and minimum.

– However, the gyratory crusher is sensitive to jamming if it is fed with a sticky or moist product loaded with fines. This inconvenience is less sensitive with a single-effect jaw crusher because mutual sliding of grinding surfaces promotes the release of a product that adheres to surfaces.

– Wear is maximum at the lower part, because at this location compression forces are exerted on the entirety of grinding surfaces.

– The profile of active surfaces could be curved and studied as a function of the product in a way to allow for work performed at a constant volume and, as a result, a higher reduction ratio that could reach 20. Inversely, at a given reduction ratio, effective streamlining could increase the capacity by 30%.

– The typical reduction ratio is bound to be between 6 and 8 and is similar that of jaw equipment.

– Applying the crushing force by rolling the nut breaks flats and gives a product that is more "cubic" than the jaw crusher.

– The largest-scale equipment can treat a flow with the magnitude of 4,000 tonne.h^{-1}.

– With a production rate and flow that are equal, the power consumed by a gyratory crusher is similar to a jaw crusher.

4.1.9.1. *Cone crusher*

It functions according to the same kinetic principles as a gyratory crusher, but instead of being suspended, the nut is supported at its base.

Compared to the gyratory crusher:

– its size and capacity are much smaller;

– its rotation speed is clearly much higher;

– the product fineness is much greater (up to 2 cm for a secondary crusher and up to 3 mm for a tertiary crusher);

– the reduction ratio varies between 6 and 8.

Profiles of the chamber and the nut are adjusted to the desired fineness.

4.2. Shock equipment

4.2.1. *Principle*

This equipment is of two types:

– hammer crushers and mills;

– cage mills.

The rotor in hammer equipment has a horizontal axle and, at its periphery, it has hammers spread on a crown that can hold 2, 4, 6 or 8 hammers. The number of crowns could, along the length of the axle, vary between 1 to over 30 (sugar cane grinder).

Hammers could be organized in relation to the rotor or rigidly fixed upon it. Rigid hammers allow for a maximum capacity and are used in crushers. Organized hammers used in grinders are recommended for easy products (limestone, clay, standard bricks, cork) and for moderate flows. Organized hammers each act separately, which allows for all piece sizes to be attacked.

Hammers are made of steel or chrome (30%) hardened by incorporating 1% manganese and 1% silicon.

The rotor and its hammers turn in a cylindrical volume that could contain:

– a localized impact plate and a plane surface;

– an impact plate (the anvil) is set out around the rotor, which covers the major portion of its perimeter.

The anvil, particularly used for crushers, is generally made up of a series of collision bars with a rectangular section, from which the product rebounds to be hit by hammers. The impact plate could also be a plate with a sawtooth profile that has the same effect. Finally, with sticky products or, products rich in fines and moisture, there is a risk of accumulation in the equipment and, finally, obstruction and jamming. The anvil is therefore made up of mobile plates mounted in a similar way to tracks on a tractor, so that the work surface is constantly being renewed.

The impact plate's bars are spaced one or several centimeters apart and let the crushed product pass without being selective to the grain size. If the anvil is a surface without holes or voids, the product is pushed back and collects at the base of the equipment. It is therefore crucial to plan for an external classifier, so that the equipment recycles the fraction of inadequate fineness.

We can avoid this complication, especially in grinders, by replacing the anvil with a perforated sheet or series of bars that will let only the adequately fine fraction escape.

The percussion effect could be achieved with a rotor spiked with "teeth" or with points of unequal size. The rotor turns in the mill chamber that gets increasingly narrow as the product advances. The peripheral speed of the "teeth" is 2 m.s^{-1}.

The percussion effect can also be achieved by the hammer's action. As such, the cage grinder has a cylindrical rotor formed by bars parallel at the rotation axle. This cage turns on the inside of another bar cage whose diameter is slightly greater.

In impact crushers and grinders, the product fills the ring-shaped gap located between the rotor and the anvil. It turns with the rotor but much slower.

The feed system for grinders and crushers is located at the upper part of the equipment that forms a hopper. Boulders to crush or grind descend into the hopper as crushing or grinding takes place. At the inlet, the product undergoes the first hard collision from a hammer and then mixes up with the product roll rotates with the hammers (but more slowly than them). Several collisions come one after the other, with a lower intensity that depends on the difference between the peripheral hammer speed and the rotation speed of the product roll.

Tschorbadjiski and Schalinus [TSC 87a, TSC 87b] have quantitatively described the product's behavior in hammer equipment.

The product's exit can take place in different ways:

– by centrifugal force through the grid surrounding the rotor and its hammers;

– in the same way between collision bars and the impact plate;

– by gravity with the lower part of the equipment if the anvil is designed without interstitial voids;

– by pneumatic drive through an exit located, like for the feed system, at the upper part of the equipment and that can be placed on the other extremity of the rotor if the machine is long and has many crowns of hammers.

Mill hammers are much smaller than hammers on crushers and there are many more of them. A large number of hammers increases not only the fineness of the product but also the energy consumed.

The shape of teeth varies with the characteristics of the ore: pointed in the case of coal, they could also be up to 20 cm high for crushing rocks. The anvil could be smooth or ribbed and powerful springs keep it in place.

A rotor with teeth could give products with an 80% underflow that does not exceed 50 μm.

4.2.2. *Collision equipment performance*

We can increase the fineness of the product in four ways:

– by increasing the rotation speed of the rotor. As such, the peripheral hammer speed is 50 m.s^{-1} for a crusher and 100 m.s^{-1} for a grinder;

– by decreasing the gap between the hammer's hitting circle and the anvil if it is set. As such, as sugarcane grinding progresses, this gap is reduced from 15 to 5 mm along the length of the rotor's axle. Nevertheless, the lower part of the equipment (toward the outlet), and the gap between the anvil and the circle of hammers or teeth cannot be decreased too much because of flow rate;

– by decreasing the opening of the outlet grid of a mill or even the opening of the screen connected to a crusher. But, as product flow decreases, the product spends a longer time in the equipment and energy consumption per tonne of the product increases;

– by increasing the number of hammers, meaning the number of circles that they are placed on.

Classifiers connected to collision mills are of cyclone type. However, a simple air current sweeping the inside of the equipment is sometimes enough to drag particles that have a satisfactory size without needing to pass through a cyclone. Classifiers connected to crushers are screens.

Large crushers can treat up to 1,500 tonne.h^{-1} of rocks with an installed power of 4,000 kW. Sugarcane mills can have a flow rate of 200 tonne.h^{-1} of cane. Equipment used for very fine grinding has a capacity that normally does not exceed 10 tonne.h^{-1}. Finally, small grinders have a production limit of 100 kg.h^{-1}.

Hammer crushers require 30% more energy compared with gyratory equipment or jaws as the kinetic energy of fragments is released as a pure loss. Hammer mills are still mainly used to get ultrafine products, for example, grain flours. In this case, the high energy consumption is from the plastic behavior of the soft product and it has nothing to do with the type of machine used.

When treating ores and inorganic products, the granulometry obtained is spread out and the portion of fines is quite significant. On the other hand, fibrous plant products like sugarcane give, after grinding, a mass of cell debris mixed with long woody fibers. In this case, the idea of granulometry does not make sense and neither does the fines content.

4.2.3. *Products treated using hammer equipment*

Wearing of hammers is still significant and it is not advised to use a hammer crusher or mill with abrasive products like coke or firebrick. In all situations, the content of silica in the treated product must not exceed 10%. This equipment is very suitable for soft products. As such, talc can be milled to a fineness of 10 μm.

Soft rocks, and therefore quite easily deformed, have a tendency to plug gyratory crushers or jaw crushers. On the other hand, with collisions caused by hammers, they appear to be more fragile and are treated in a satisfactory way by using hammer crushers.

Hammer mills do not accept slurries.

Generally speaking, if an ore or plant's moisture content is too high, the product tends to exhibit plastic behavior and grinding it becomes more difficult. In particular, if the feed is enough rich in fines and moisture so as to be sticky, it will have a tendency to accumulate on the equipment, a solution to which would be to add dry product to the feed. Drying by sweeping hot air is required if the plan is to use a pneumatic classifier.

Hammer mills allow for a good release of the main constituents of an ore, because with collisions pieces literally explode and the ore is liberated from its matrix. In fact, matrix–ore interfaces are weak surfaces.

The hammer crusher works quite well as the primary crusher on soft rocks or moderately hard ones (schistose and stratified structures) and the reduction ratio could be approximately 4. If working in a "closed circuit" (so with one classifier) and if treating limestone (which is neither hard nor abrasive), the hammer crusher could reduce 500 mm boulders to a size that is less than 15 mm in one stage, which simplifies the crushing pattern.

This equipment is very suitable for treating products with a flat shape. The crushed product has a shape that is similar to a cube. It could be suitable for moist and sticky products.

4.2.4. Modeling hammer and anvil mills

These machines work in a closed circuit with an external hydraulic classifier that sufficiently evacuates fine products and recycles any product that is too big back into the mill.

Air circulation in the interior of the mill creates an internal pseudo-classification, whose effect is additional to the external classifier. We will only study the effect of internal classification, that is the operation in an open circuit where production is expelled pneumatically.

See Figure 4.3, where, f_i, f_i', m_i and p_i are the specific fractions with a size x_i, respectively, in the fresh feed, at the inlet of a proper grinding operation, inside the mill and in the product that is exiting.

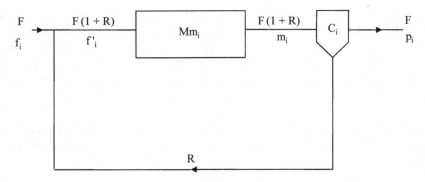

Figure 4.3. *Internal classification circuit*

F will be the total feed flow (which is equal to production flow) and M is product holdup inside the machine.

C_i are all elements of the diagonal matrix that characterizes the classifier.

F is measured in $kg.s^{-1}$ and M in kg.

For these detailed calculations, we draw heavily on the results of:

– Austin and Gotsis [AUS 79];

– Austin *et al.* [AUS 80];

– Rogers [ROG 82];

– Austin and Luckie [AUS 82a, AUS 82b].

The balance of the loop is written as:

$$F(1+R)f_i' = Ff_i + F(1+R)C_im_i \qquad\qquad [4.1]$$

$$\left(\text{with: } R = (1+R)\sum_1^n C_im_i \right)$$

Milling in the equipment is expressed by:

$$F(1+R)m_i = F(1+R)f_i' - S_im_iM + \sum_{j=1}^{i-1} b_{ij}S_jm_jM \qquad\qquad [4.2]$$

Eliminating f_i' between the two equations [4.1] and [4.2], we get:

$$F(1+R)m_i = Ff_i + F(1+R)C_im_i - S_im_iM + \sum_{j=1}^{i-1} b_{ij}S_jm_jM$$

Let us set:

$$\tau = M/(1+R)F \quad \text{and} \quad m_i^* = (1+R)m_i$$

Let us divide by F:

$$m_i^* = f_i + C_im_i^* - S_im_i^*\tau + \sum_{j=1}^{i-1} b_{ij}S_jm_j\tau$$

Where:

$$m_i^* = \frac{f_i + \tau \sum_{j=1}^{i-1} b_{ij} S_j m_j^*}{\tau S_i + 1 - C_i}$$ [4.3]

Now let:

$$\gamma_i = \frac{M}{F(1+R)} S_i m_i (1+R) = \tau S_i m_i^*$$

Equation [4.3] is written as:

$$\gamma_i = \frac{f_i + \sum_{j=1}^{i-1} b_{ij} \gamma_j}{1 + \frac{1 - C_i}{S_i \tau}}$$ [4.4]

Components of the granulometric vector of the product are:

$$p_i = r m_i M (1 - C_i)$$ [4.5]

where r is measured in s^{-1} and it characterizes the fraction by mass pulled by the gas current.

On the other hand, we see on the circuit graph that:

$$p_i = F(1+R) m_i (1 - C_i)$$ [4.6]

Where, by eliminating p_i between [4.5] and [4.6], we obtain:

$$r = \frac{F(1+R)}{M} = \frac{1}{\tau}$$

Equation [4.4] is then written as:

$$\gamma_i = \frac{f_i \sum_{j=1}^{i-1} b_{ij}\gamma_j}{1 + r\frac{(1-C_i)}{S_i}}$$

γ_i are all calculated by recurrence starting from γ_1 if we are given b_{ij}, S_i, r and C_i.

Equation [4.4] becomes:

$$Mm_i = \frac{\gamma_i}{S_i} = \frac{f_i + \sum_{j=1}^{i-1} b_{ij}\gamma_j}{S_i + r(1-C_i)} \qquad [4.7]$$

4.2.5. Determining r and C_i experimentally

We use equation [4.6]:

$$p_i = r(1-C_i)m_i M$$

We carry out the granulometric analysis of the contents of the mill after stopping it, which provides values for m_i as well as M. Furthermore, we analyze the product at the outlet to obtain values for p_i.

For fines, we state that $C_i = 0$ such that:

$$r = \frac{p_i}{Mm_i}$$

Knowing r, we deduce C_i from the most significant sizes and express them in the following way:

$$C_i = \frac{1}{1 + (d_{50}/x_i)^\lambda}$$

We also note that:

$$(1+R) = rM; \text{hence the R value}$$

We could also have used the relation:

$$R = (1+R)\sum_{1}^{n} C_i m_i$$

4.2.6. *Determining milling parameters*

If we make use of analytical expressions for B_{ij} and S_j, we calculate the corresponding parameters with a gradient method using equation [4.7] to compare calculated m_i values with experimental m_i values.

The $S_i M$ product increases with retention and approaches a limit. It is the same for net power.

4.2.7. *Modeling grid equipment*

In these devices, comminution is a result of the product's impact on hammers that are fixed in relation to the rotor.

The continuous grinding equation applies to this equipment:

$$p_i = f_i - S_i m_i \tau + \tau \sum_{j=1}^{i-1} b_{ij} S_j m_j \qquad [4.8]$$

Remember that S_i and S_j are measured in s^{-1}.

τ: the average residence time:

$$\tau = M/F$$

M: retention of solid in the equipment (kg)

F: fed flow (kg.s^{-1})

The flow that passes through the grid is equal to the fed flow that is found integrally in the product:

$$Fp_i = r(1-g_i)m_i M$$

or even:

$$p_i = r(1-g_i)\tau m_i \tag{4.9}$$

In the above equation, r is measured in s^{-1}.

It can be noted that g_i characterizes the grid's power classifier for each granulometric layer. The grid's opening is such that if the granulometric layer with the index i is present at Mm_i kg inside the equipment, the fraction $r(1-g_i)$ passes through the grid in 1 s.

Eliminating p_i between both equations [4.8] and [4.9]:

$$r(1-g_i)m_i\tau = f_i - S_i m_i\tau + \tau\sum_{j=1}^{i-1}b_{ij}S_j m_j$$

Let:

$$\gamma_i = S_i m_i$$

The preceding equation becomes:

$$\gamma_i = \frac{f_i + \sum_{j=1}^{i-1}b_{ij}\gamma_j}{1+\dfrac{r(1-g_i)}{S_i}} \tag{4.10}$$

γ_i are all calculated successively from γ_1.

Equation [4.8] becomes:

$$p_i = f_i - \gamma_i + \sum_{j=1}^{i-1} b_{ij} \gamma_j \qquad [4.11]$$

4.2.8. Determining parameters

By stopping the mill, we can measure the mass M of the retained solid and the granulometry m_i.

Knowing the feed flow rate F, we deduce the average residence time:

$\tau = M/F$

The granulometry p_i of the product obtained during normal operation is measured in the laboratory.

Equation [4.9] then allows us to calculate:

$r(1 - g_i)$

This product is a decreasing function of the size of particles and, when the index i changes from 1 to n, the quantity r $(1-g_i)$ increases from zero to a few s^{-1}.

If we make use of analytical expressions for B_{ij} and S_j, we can then calculate the corresponding parameters with a gradient method using equation [4.11] and by comparing calculated p_i values with experimental p_i values.

By increasing the feed flow F, we see that M changes in the same way until there is a jam.

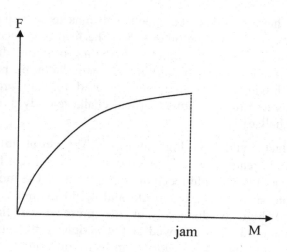

Figure 4.4. *Relation feed – retention*

In the same way, the function $M_r (1-S_i)$ grows with M.

The product $S_i M$ changes with M in the same way as the feed flow rate F.

Finally, the useful power is proportional to the feed F above a certain minimum value of the holdup M.

Jindal and Austin [JIN 76], Austin *et al.* [AUS 79] and finally Gotsis *et al.* [GOT 85] have experimentally proven the above results.

4.2.9. *Shredder*

A shredder is made up of a rotation axle with what we can call teeth fixed up on it. This equipment is used to shred used tires and cotton or hay.

4.2.10. *Corrugated rotor mills and pin mills*

– For very fine grinding (< 10 μm) of dry plant material, we use a device made up of a cylindrical rotor turning at 3,000 rev.mn^{-1} in a stator with a slightly bigger diameter. Rotor and stator surfaces are equipped with parallel corrugations along the vertical axle. Fines are pulled toward the top.

– Pin mills have two discs equipped with pins that are parallel to their axle. One disk is fixed and the other turns. The feed is located at the center of the fixed disc. Pins are set in circles alternating from one disk to another. Fragmentation is carried out by pulling out and shattering pieces. Fibrous products are shredded. This equipment is used for soft products with a hardness that is less than 3 Mohs (carbon, chalk, talc, dyes, drugs, pepper, sugar, resins, shellac).

– The product could have high moisture. The peripheral speed of the turning disc could reach 120 $m.s^{-1}$. It is possible to achieve a fineness that is less than 5 µm and flow could reach up to 5 tonne.h^{-1}. The mobile disc turns at an adjustable speed of between 700 and 4,500 rev.mn^{-1}. A high speed corresponds to high production, but the presence of fines could be significant. The product is removed at the periphery by centrifugal force. This equipment falls into the category of mills for ultrafines.

4.3. Roller crushers and mills

4.3.1. Operation principle

Two cylinders along the parallel and horizontal axles turn in the opposite direction, very close to each other at equal or similar speeds. The product is generally fed by gravity above the gap separating the two cylinders.

4.3.2. Angle of nip

Let us consider at a given moment a spherical particle that is pinned between two cylinders. It is subjected in part to the surface of each cylinder by a normal force N and a tangential exertion T. The particle's weight is negligible compared with these forces. The *angle of nip* θ is the angle between the planes tangent to the cylinders at the particle's two points of contact, so that there is not any sliding, and it must be that:

$$T < T_{max} = \mu N$$

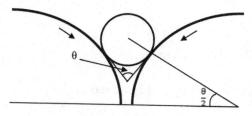

Figure 4.5. *Angle of nip*

The solid friction coefficient μ has a magnitude of 0.15 for wet products and 0.3 for dry products.

For the particle not to be released, you must have the vertical component T (pointing toward the bottom and equal to $T\cos\theta/2$) being greater than the vertical component of N (pointing toward the to and equal to $N\sin\theta/2$).

In other words:

$$T > N \, tg\frac{\theta}{2}$$

For this to be possible, you must have:

$$T_{max} \geq N \, tg\frac{\theta}{2}$$

or even:

$$\theta < 2 \, Arc \, tg\mu \qquad\qquad [4.12]$$

Taking into account the variations in μ, θ must range between 17° and 33°.

If x_F is the particle diameter and e the minimum gap separating the two cylinder's surfaces, we find that:

$$\cos\frac{\theta}{2} = \frac{D+e}{D+x_F}$$

where D is the diameter of the cylinders.

Let:

$$\alpha = \cos\frac{\theta_{max}}{2} = \cos\left(\text{Arc tg}\mu\right)$$

According to the inequality [4.12] we have:

$$\frac{D+e}{D+x_F} > \alpha$$

This result enables us to define the maximum reduction ratio R_M that the machine is capable of for a given product:

$$R_M = \frac{1}{\alpha}\left[\frac{D}{e}(1-\alpha)+1\right]$$

This reduction ratio is optimal for a dry ore. With large cylinders and significant spacing, crushers could treat 80 mm pieces with a reduction ratio of 4–7.

The idea of an angle of nip comes into play especially for crushers and sometimes for mills. In fact, in the latter, the diameter of the rollers is never less than 0.2 m.

NOTE.–

Johanson [JOH 65] proposed a theory for crushing between two cylinders, but it does not go so far as to express mill behavior.

EXAMPLE.–

$$\mu = 0.3 \quad D = 0.5 \text{ m} \quad e = 4.10^{-3} \text{ m}$$

$$\frac{\theta_{max}}{2} = \text{Arc tg}0.3 = 16.7°$$

$$\alpha = \cos16.7° = 0.9578$$

$$x_F < \frac{0.5(1-0.9578)+0.004}{0.9578}$$

$$x_F < 0.026 \text{ m}$$

$$R_M = \frac{0.026}{0.004} = 6.5$$

4.3.3. *Sliding between cylinders and shearing*

Sliding and shearing occurs between cylinders when the absolute values of the peripheral speeds are different.

When sliding is nil, comminution is carried out by pure compression. The energy yield is then important but it decreases with the reduction ratio, as there is agglomeration of fragments and loss from mutual friction.

Sliding of 5–10% produces a shearing effect that is useful for treating plastic products. This is the case for paint mills. The product is then pulled by the fastest cylinder. These machines, for mechanical reasons, are made up of three cylinders and therefore undergo two successive grindings. Such sliding is not advised for products with dry grains, as there will be a significant production of fines.

However, if the ratio of peripheral speeds achieves a value of 10, the energy yield increases again as significant shearing prevents fragments from agglomerating. This yield could even be greater than the value that it would have with no sliding.

4.3.4. *Roller equipment capacity*

Capacity is given by the equation:

$$W = F(1-\varepsilon)\rho_s e \overline{V}_p L \ \left(kg.s^{-1}\right)$$

F: empirical factor that increases when the angle of nip decreases

ε: product's porosity between cylinders. If the product is soft, there will be a tendency to agglomerate and ε will be low and clearly less than 0.4

L: length of cylinders (m)

$\overline{V}_p$ rmean peripheral speeds of two cylinders (m.s^{-1})

e: gap between two cylinders (m)

ρ_s: true density of product (kg.m^{-3})

EXAMPLE.–

We grind wet anthracite:

$$\epsilon = 0.3 \quad F = 0.6 \quad \rho_s = 1,600 \text{ kg.m}^{-3}$$

$$L = 2 \text{ m} \quad e = 3.10^{-3} \text{ m} \quad \overline{V}_p = 8 \text{ m.s}^{-1}$$

$$W = 0.6 \times (1 - 0.3) \times 1600 \times 3.10^{-3} \times 8 \times 2$$

$$W = 32.2 \text{ kg.s}^{-1} = 116 \text{ tonne.h}^{-1}$$

4.3.5. *Fragmentation rate*

We will not use the classical notion of fragmentation speed, here but instead a fragmentation rate which is a dimensionless number less than one. We will denote this rate by a lowercase "s" to distinguish it from the fragmentation speed S_j used for most equipment.

When grains in the granulometric layer j are found between two rollers, the fraction $(1-s_j)$ is not fragmented at first while the fraction s_j undergoes fragmentation. At the interior of this fraction s_j, a fraction s'_j is fragmented a second time and the proportion $(1-s'_j)$ is not fragmented again.

We characterize the granulometric layer using the index i_g, whose average grain size x_i is equal to the spacing between rollers.

If $i \geq i_g$, grains that are too small to be crushed directly are fragmented by compresstion of adjacent grains, in both step s and step s'. We will write:

$$s'_i = s_i$$

If $i < i_g-1$, grains are, in all situations, crushed by the rollers during step s, the output is fragmented again during step s'. This step s' acts on grains that are already much smaller than x_i, but always by crushing between the rollers. Then, we will write:

$$s'_i = s_{i-1}$$

If $i = i_g - 1$, there is a granulometric transition layer and we simply write:

$$s'_i = \frac{1}{2}(s_i + s_{i-1}) = \frac{1}{2}\left(s_{i_g-1} + s_{i_g-2}\right)$$

We then very easily obtain values for s'_i from values for s_i. In addition, the value of s_i increases with the grain size, because the bigger the grain, the greater the chance it will be reduced, even if it has the shape of a needle. Therefore:

$$s_{i-1} > s_i$$

On the other hand, s'_i is greater than s_i because a grain of size x_i that escaped during the first fragmentation s_i is found to be more in the front in the gap between rollers and, at this level, cylinder separation is low. The grain therefore has a greater probability of further fragmentation. In other words:

$$s'_i > s_i$$

Expressing s_i remains and this is done in the following way:

$$1 - s_i = \frac{1}{1 + \left(\dfrac{x_i}{x_{50}}\right)^{\lambda}}$$

Components of the fragmentation rate s_i are measured experimentally by determining the fraction of a unique feed size x_i that is not fragmented when passing through the gap separating the rollers. This fraction is:

$$1 - s_i + s_i\left(1 - s'_i\right)$$

4.3.6. *Modeling equipment with smooth cylinders*

Figure 4.6 illustrates previous developments.

We admit that fragmentation of each granulometric layer is done independently of other layers. We take up again a part of the calculations made by Austin *et al.* [AUS 80, AUS 81b, AUS 81c] by modifying them for increased precision.

Consider then that we feed the equipment with a unique layer with a size x_j with a flow f_j.

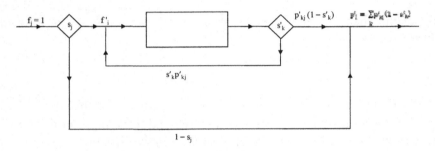

Figure 4.6. *Modeling the roller mill*

Selecting feed particles that get milled could be materialized by a fictitious separator s_j.

On the other hand, the existence of the equally fictitious separator s'_k sends the fraction of grains at the outlet that undergo a second milling.

We observe at the outlet:

$$p_{kj} = p'_{kj}\left(1 - s'_k\right) \qquad j+1 \le k \le n \qquad [4.13]$$

p_{kj} is the mass of product size x_k coming from feed size x_j.

Note that, for $k = n$

$$s_n = 1 \quad \text{and} \quad s'_n = 0$$

The outlet p'_{kj} of a mill size x_k includes:

– fragments coming directly from the grinding of f_j:

$$b_{kj}s_jf_j$$

– the "recycled" p'_{ij} that corresponds to sizes i absolutely lower than j (where $i \geq j + 1$). Besides, these sizes, to give size fragments x_k in the product, must be such that $x_i \geq x_k$, meaning $i \leq k$. But what is recycled by the separator s' was by definition completely fragmented in the mill. All that remains in the recycled fraction is grain size x_k. In other words:

$$b_{kk} = 0$$

Recycled granulometric layers therefore corresponds to i values that do not exceed the value $k - 1$.

Finally:

$$j + 1 \leq i \leq k - 1$$

Austin *et al.* [AUS 81c] deduced that:

$$b_{kj} = \frac{\dfrac{p_{kj}}{1 - s'_k} - \sum_{i=j+1}^{k-1} \dfrac{b_{ki}s'_i p_{ij}}{(1 - s'_i)}}{s_j f_j} \qquad [4.14]$$

with

$$j + 1 < k \leq n$$

We can calculate b_{kj} by recurrence by admitting that the distribution matrix is standardized:

$$b_{kj} = b_{(k-j)} = b_q$$

We will start with:

$$b_{j+1,j} = b_1 = \frac{p_{j+1,j}}{(1 - s'_{j+1})}$$

This calculation assumes that we have values for s_j, s'_j and p_{kj}. Yet, we have seen that s_j is measurable and that s'_j is determined by the calculation. p_{kj} must be measured. Rogers [ROG 82] suggests an explicit method for calculating the terms d_{ij} of the mill's transfer function.

Finally, for a feed with n granulometric layers, the mass of the layer of product size x_k is:

$$p_k = \sum_{j=1}^{n} p_{kj}$$

and, of course:

$$\sum_{j=1}^{n} f_j = F = P = \sum_{k=1}^{n} p_k$$

4.3.7. *Operational and performance conditions*

The peripheral cylinder speed could be much higher for wet products $(8–10 \text{ m.s}^{-1})$ than for dry products $(4–6 \text{ m.s}^{-1})$ because there is less heating of the cylinders.

Crusher cylinder separation is never less than 1 mm. On the other hand, in flour milling, we use smooth cylinder mills called sizing rolls and converters for the finest fractions. Cylinder separation is then less than a millimeter.

On this basis, oleaginous grains are also milled in smooth cylinder mills and we get fine scales with a thickness that is less than a millimeter, which improves oil extraction either by pressing, or by solvent extraction. This type of milling that produces flakes is called flattening or lamination.

The reduction ratio of crushers and mills is between 3 and 7 and its magnitude is often 4.

The granulometry obtained will be constricted if we distribute the feed evenly and with a low flow rate along the length of cylinders. The target reduction ratio will then be approximately 3. Conversely, if there is no fear of over-milling and if we can allow for a spread granulometry, the feed could accumulate on cylinders. The reduction ratio will then be much higher because grains grind among themselves by erosion.

The major inconvenience with cylinder equipment is wear and this limits their usage for products like coal or even food products. Hard minerals are excluded from the product's Mohs hardness, which must be less than or equal to 5 (clay, shale, ash, barite).

The industrial yield of cylinder equipment is the best (60–90%) of all crushers and mills.

4.3.8. *Characteristics of products at outlet [GAU 26]*

Kellerwessel [KEL 86] suggested with his roller press, prior compaction before proper grinding. According to the author, production is thus increased and energy consumption is reduced.

The roller press must be fed quite slowly for a volume of product to accumulate above the "gap" of the two rollers. Particles are crushed by mutual contact, which allows hard products such as quartz to be treated. Energy consumption is less than 3.5 kWh.tonne^{-1}.

In the small particle domain, we obtain the straight line AB represented by $\log_{10}m_i = f(\log_{10}x_i)$ (see Figure 4.7).

The steep slope A'B shows a decrease in the reduction ratio and the presence of less fines. A feed with flat grains gives a much lower slope than a unidimensional feed. Cubic grains break in two ways. First, transverse cracks make tablets, especially if the reduction ratio is low. Then, radiating cracks appear near the compression force's point of application. The biggest grains are fairly flat but those the peak M of the frequency curve are more unidimensional. The finest grains have a tendency to be lengthened, or at least angular.

It is clear that galena (soft) produces more fines than hard quartz for the same adjustment of the separation gap between rollers at the size of the feed.

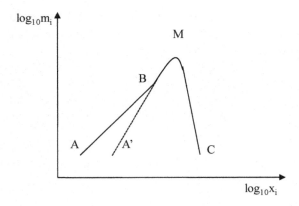

Figure 4.7. *Frequency curve at the roller press outlet*

4.3.9. *Equipment with grooved rollers*

When a product has small independent grains but is somewhat plastic, we can expect to find grooves on cylinders that are shifted from one cylinder to another, which leads to a shearing effect that improves fracture.

Roasted coffee grains are ground with equipment whose cylinders have grooves that are parallel to the axle.

In flour production, grains of wheat are first treated (after hydrating to 16 or 17% moisture) with cylinders equipped with helical slots, with very broad gaps and triangular sections. From the first to the last couple of cylinders:

– the section of grooves decreases;

– the number of grooves increases from 300 to 800 for cylinders that are 0.2 m in diameter;

– gaps between cylinders decrease.

Sliding is expected to emphasize the shearing effect. Milling is then followed with machines with smooth cylinders.

Typically, mills with grooved cylinders are used to carry out the initial milling of grains.

4.4. Track crushers and mills

4.4.1. *Equipment without a built-in classifier*

Two grindstones roll in a chamber. These grindstones are made of steel or nickel. The vertical pressure force exerted by springs on the grindstones could reach 20 times but sometimes we are only dealing with the weight of the grindstones.

A grid is located at the bottom of the chamber, with openings whose size varies from 1.5 to 10 mm. We can also expect slots along the chamber's periphery.

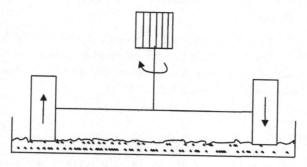

Figure 4.8. *Grindstone mill*

This equipment is suitable for moderately hard products (clay, schist, barite). It consumes low power and requires very little maintenance because it is a simple machine.

The reduction ratio is high; from pieces that are 10 cm in size we can obtain grains ranging in size from 1 to 5 mm.

The maximum flow is 50 tonne.h^{-1}.

A wet path could be used, particularly for ceramic powders.

The olive grinder used to produce olive oil functions by using the same principle.

4.4.2. *Equipment with built-in classifiers*

The classifier used is always of the hydraulic type.

The track is a ring with an immobile vertical axle. The axle bearings that are slightly inclined on the vertical are applied to the track by centrifugal force. This is therefore a swing-type mill. It is suitable for soft products and allows for fines that are 0.1 mm. The potential flow rate does not exceed 50 tonne.h^{-1}.

For coal, we could use either:

– equipment with turning chambers on the edges of which bearings are applied using springs;

– or a system of steel balls sealed between two circular tracks where one is fixed and the other is revolving. Springs exert a vertical face on the fixed upper track. The centrifugal force removes the product. This model was modeled by Austin *et al.* [AUS 81b, AUS 81c].

4.4.3. *Net power*

Remember that net power P_N is the difference between the power required for normal equipment function and the power when empty (without product).

Now let F_B be the exertion of grinding bodies on the track. By definition of the friction coefficient μ, the product μF_B is the orthoradial resistance force that is opposite to the relative movement of grinding bodies and the track. The corresponding torque is:

$$C = R\mu F_B$$

where R is the radius of the circular track.

The net power is then:

$$P_N = C\omega = \mu F_B R\omega$$

The coefficient μ varies with the nature and granulometry of the treated product. Its value has a magnitude of 0.1.

To sufficiently fill the track with the product, the expression $S_i\,M_p/P_N$ is independent of the mass M_p of the product on the track. S_i is the fragmentation speed relative to the granulometric layer size x_i.

4.4.4. Range of similar mills

In the manufacturer's range, equipment is geometrically similar.

In machines where the milled product is evacuated by centrifugal force, rotation speed is such that, at the edge of the track, the centrifugal acceleration is independent of the track's radius:

$$\omega^2 R = \text{const.}$$

On the other hand, the pressure of grinding bodies supposedly spread on the track is also constant.

$$F_B/(2\pi RL) = \text{const.}$$

Yet, the width L of the track is proportional to R. So:

$$F_B \sim R^2$$

Net power is such that:

$$P_N \sim F_B R\omega \sim \frac{R^3}{\sqrt{R}} \sim R^{2,5}$$

It is the same as the flow that is proportional to net power.

4.4.5. Detailed study of ball-and-race (track) mills

Broadbent and Calcott [BRO 57] study the open circuit cone mill. This mill has a cone with its tip on top and whose axle is that of the equipment. The cone is located under the feed. These authors give a diagram corresponding to Figure 1 in their publication. In the same publication, they describe the closed circuit mill operation, meaning with a classifier.

Austin *et al.* [AUS 81b, AUS 81c] study milling kinetics, meaning the breaking speed and terms of the fragmentation matrix on a hardgrove-type coal mill. The Babcock E 1.7 mill shown in a cross-section in Figure 1 of Austin *et al.* [AUS 82a] is the same type of mill. These two types of mills are compared by the same authors [AUS 82a].

Expressions are suggested respectively for the capacity and power consumed by the same authors [AUS 82b] for a ball-and-race mill on which steel balls circulate.

NOTE.–

Ball-and-race mills are only used for dry milling.

NOTE.–

In the Hardinge mill, grinding bodies are big boulders that circulate along the circular horizontal track. The curve for the distribution frequency of diameters is a unique maximum flanked by two straight lines.

4.5. Autogenous mill

4.5.1. *Principle*

Rod or bar mills suffer from a lot of wear if the product is very abrasive. This limitation gave rise to the autogenous mill where the grinding bodies are themselves pieces of ore. These pieces are a lot more efficient when they fall from a great height. This is why the diameter of these mills could achieve 11 m and their length only 4.5 m. They could require an installed power of up to 9,000 kW.

Internal retention in the mill basically involves:

– large-sized boulders;

– particles (grains) that have a small size or an average size.

Comminution is performed:

– by crushing grains by collision, either mutually or with boulders;

– by erosion, meaning detaching fine particles from average-sized grain surfaces or from boulder surfaces. As such, the production of fines could be significant, which could be an inconvenience (but not always).

Generally, ore density is much less than the density of steel. To activate fragmentation, we expect to have on the inside of the equipment a low proportion (5% by volume, for example) of steel balls. Wear on this low mass of balls has very little effect on the actual operation cost. Mills that operate in this way are called semi-autogenous mills.

The feed must have a suitable granulometry. In particular, there needs to be an adequate proportion of boulder sizes greater than 250 mm. The shortage of large boulders must be compensated by the addition of balls and operation in a semi-autogenous mill.

The rotation speed is equal to 85 or 90% of the critical speed. Like ball mills, the critical speed is given by:

$$N_c = \frac{42.3}{\sqrt{D}} \ \left(\text{rev.mn}^{-1}\right)$$

D: equipment diameter (m).

4.5.2. Modeling

Stanley [STA 74] uses the equation for continuous milling which, in a matrix form, is written as (where (p) and (f) are flows):

$$(p) = (f) + [b][S](m) - [S](m) \tag{4.15}$$

If [g] is the matrix of the underflow's coefficients through the outlet grid, we thus have:

$$(p) = \frac{1}{\tau}[g](m) \tag{4.16}$$

Where [g] is the diagonal matrix and, if the grid is ideal, elements g_i of the matrix are equal to zero when $i \leq g$ and equal to 1 when $g < i \leq n$. On the other hand, τ can reach 1.5 mn.

By substituting (p) in [4.14], [4.15] and [4.16], we obtain:

$$(f) = \left(\frac{[g]}{\tau} + [S] - [b][S] \right)(m)$$

and, taking into account [4.16]:

$$(f) = \left(\frac{[g]}{\tau} + [S] - [b][S] \right)\tau[g]^{-1}(p)$$

This relationship directly relates the feed to production.

Elements $S_j(x_j)$ of the fragmentation speed matrix give two maxima as a function of x_j. The peak corresponding to the lowest values of x_j deals with crushing grains, and the other peak with boulder abrasion. The analytical expression for S_j is very empirical and its value is between 5 and 100 h^{-1}.

As a function of grain size in the device, the fragmentation function can be represented in *three different forms*:

1) j is small (boulders): erosion dominates almost exclusively. We use a non-standardized function:

$$B_{ij}^A = 1 - \exp\left[-\left(\frac{x_i}{x^*} \right)^u \right]$$

This expression involves empirical parameters x* and u.

2) j is big: it is the fine fraction where crushing is the main process:

$$B_{ij}^E = E\left\{ 1 - \exp\left[-\left(\frac{x_i}{x_j} \right)^v \right] \right\}$$

This function, termed Broadbent, is standardized.

3) j is medium: we then have

$$B_{i,j}^M = \varnothing_j B_{ij}^A + \left(1 - \varnothing_j \right) B_{ij}^E$$

This zone corresponds to a variation from approximately 1 to 6 in grain size. To locate the average in the range of grain sizes, Stanley [STA 74] distinguishes abrasion from crushing by visually examining fragments. Crushing produces conchoidal fractures while abrasion produces rounded fragments.

The mass M of solid retained in the mill at the granulometry shows two peaks if we consider the frequencies m_i (by mass) as a function of grain size. The peak that corresponds to boulders is 10 times more significant than the one for grains (see Stanley [STA 74, pp. 89 and 90]).

As such, Stanley [STA 74 pp. 81] is an example of the breaking matrix that is a combination of three matrices: erosion, crushing and intermediary. But he goes much further and replaces the erosion matrix with a combination of two matrices where, respectively:

– erosion is proportional to the particle surface;

– erosion is proportional to the particle volume.

NOTE.–

Menacho [MEN 86] provides the distribution frequency for the product exiting the mill (his equation (21)). It is limited to adding the erosion and breaking effects. He calls τ the average residence time of the powder or of the pulp in the mill.

4.5.3. Interest in autogenous mills

For an equal production and having a large equipment diameter, energy consumed is greater than that required in ball mills. On the other hand, the consumption of steel from wear on the few balls present is less but the wear on bearings is greater.

The autogenous mill could be fed with boulders reaching up to 800 mm and give, in one single operation, a product size that does not exceed a few tens of millimeters (the reduction ratio could exceed 1,000).

The possible flow increases with the diameter D of the equipment as $D^{2.82}$ and energy consumption as $D^{2.62}$ only. This explains the tendency for these machines to be gigantic as they can treat up to 300 tonne.h^{-1}.

According to its design, equipment can be used to work in a dry mode or even in a wet mode.

The autogenous mill is equipped with a classification device. In wet mode, the pulp mixture is evacuated through a grid and then treated by a hydro-cyclone or a sieve. Granules are recycled back into the equipment. In dry mode, an air current goes through the equipment and drives the fines fraction, which economizes the entire classification circuit.

The autogenous mill simplifies a circuit by replacing it with a single complete installation for crushing and milling:

– a classic diagram will include the primary gyratory, a secondary crusher in a closed circuit, two rod mills and two ball mills;

– a much simpler diagram involves only two primary crushers, two autogenous mills and one ball mill in a closed circuit.

4.6. The bead mill for ultrafines (micronizer)

4.6.1. *Description*

The equipment principle is to energetically agitate the medium in which the following are mixed:

– a liquid (generally water);

– grains of the solid to be milled, up to the size of a micron;

– small beads that act as grinding bodies and whose diameter has a magnitude of a millimeter.

In all configurations, agitation is achieved by a turning device whose vertical rotation axle is shared with a cylindrical enclosure called a chamber. Different types are obtained by sharing the chamber volume between the agitation system on the one side and the agitated medium on the other side. The types of chambers are described by Engels [ENG 87]) and Schütte *et al.* [SCH 83].

If the agitated volume is the maximum and represents, for example, 80% of the chamber's volume, the rotor's shaft has a diameter that is simply enough to make sure that it has enough mechanical resistance and that it can carry full or open pallets or discs [BEC 01, STE 83, KUL 87, JOO 96a, JOO 96b, BEC 99, MEN 03]. The most common type of vertical axle is the one that is open to the upper part. The product in the water suspension is introduced to the lower part and comes out to the top through a grid that retains the grinding beads. This design is called a full chamber.

The orthoradial speed being minimum when close to the axle, this area has a very low efficiency. There is therefore interest that the product is localized close to the wall of the chamber. Therefore, we then give the rotor the shape of a simple cylinder whose diameter is such that the agitated volume is not more than 40–60% of the chamber's volume. This is the ring design [KOL 93]. This design allows for efficient cooling as the rotor could be cooled internally by circulating water (just like the chamber's wall); the surface given by the rotor for cooling the agitated medium is significant and, as the rotor can turn at a higher speed (up to 12 m.s^{-1} for peripheral speed), the thermal transfer coefficient is excellent.

4.6.2. *Energy of the micronizer*

Power consumed by the erosion mill could be expressed in the following way [SAD 75]:

$$P = KN^3D^5 \left(\rho_M + \rho_o \right) \ (kW)$$

N: rotation speed of the rotor (rev.s^{-1})

D: internal diameter of the chamber (m)

ρ_M: density of the agitated medium (liquid + product + beads)

ρ_o: constant

ρ_M and ρ_o are expressed in kg.m^{-3}.

For soft products like limestone, coal or cocoa, the energy required does not exceed 100 kW.tonne^{-1}. It could be 5–10 times higher for pigments and dyes and can even exceed 1,000 kW.tonne^{-1} for some herbicides.

While normal agitation of a liquid mixture exerts a volumic power with a magnitude of 1 kW.m^{-3}, it is very different in an erosion mill where the required power could fall between 50 and 100 kW.m^{-3} in a full chamber and even 300 kW.m^{-3} in the ring version. The result is that cooling equipment is absolutely necessary and that the industrial yield for erosion mills is less than 5%.

To model the energy spent according to the product, we usually act on the rotation speed of the rotor. If agitation is caused by shaft holding discs, the peripheral speed of the latter has a magnitude of 10 m.s^{-1}.

4.6.3. Grinding bodies (quantity in the pulp)

Beads are generally made of glass, ceramic or steel. To treat hard products like corundum, beads must be made from a very hard material. We can find potential materials in Table 2 of Joost and Schwedes [JOO 96a, JOO 96b]. Milling efficiency increases with their density. Bead size can range from 0.5 to 2 mm, increasing for products difficult to grind and can reach 6 mm for elastic products.

The concentration of beads in the medium is a very important factor for the efficiency of the operation. According to Sadler *et al.* [SAD 75], the notion of a free average path is the criterion that must be retained. According to the classic expression of the kinetic theory of gases, this path is:

$$\lambda = \frac{1}{\pi\sqrt{2}d_p^2 C_p}$$

D_B: ball diameter (m)

C_p: number of pearls per cubic meter of agitated medium

The reduced path is then:

$$\frac{\lambda}{d_p} = \frac{1}{6\sqrt{2}} \cdot \frac{6}{\pi d_p^3} \cdot \frac{1}{C_p} = \frac{1}{6\sqrt{2}\varnothing_p}$$

$\varnothing_p$: density fraction occupied by pearls in the pulp mixture.

Efficiency will be at a maximum if (see Sadler *et al.* [SAD 75]):

$$\frac{\lambda}{d_p} = 0.7 = \frac{1}{6\sqrt{2}\varnothing_p}$$

where:

$$\varnothing_p = 0.17$$

Granulometry (size distribution) follows:

Heim *et al.* [HEI 85] have shown that the equation for bulk milling expresses the granulometry of a product obtained very well, on the condition that:

$$S_i = a\left(x_i\right)^\alpha \exp\left(\Psi x_i\right)$$

$$B_{ij} = \left(\frac{x_i}{x_j}\right)^\gamma$$

Continuous milling does not appear to have been treated.

4.6.4. *Performances and use of the pulverizer*

According to the size of the equipment and the nature of the product to mill, flow rate can vary between 50 and 1,000 kg.h^{-1}.

This equipment is used to obtain ultrafine products (between 50 nm and 5 μm). It is obvious that the grinding time in batch operation increases with the desired fineness, which needs a dozen hours to grind alumina to a fineness of 40 nm. However, alumina is a hard and abrasive product.

In reality, products treated economically are soft products whose Mohs hardness is less than 5. But otherwise pearls are quickly eroded.

Particularly soft products are microorganisms (see Schütte *et al.* [SCH 83]) that are destroyed in the pulverizer in order to liberate the components.

4.7. Other equipment for ultrafines

4.7.1. *Gas jet mill*

Up until recently, a solid–gas suspension was fed directly into the grinding chamber, which gave two problems:

– injection nozzles were worn very quickly;

– product granulometry depended on its feed flow rate.

In the opposite case, for example, in equipment proposed by the Alpine Company, the product is separately fed by a screw starting from the hopper and is kept in a fluid state by the gas that is introduced by horizontal nozzles placed in circles and blowing toward the chamber's axle. We can use:

– compressed air at room temperature and between 6 and 12 rel. bar;

– hot air;

– superheated vapor.

Hot fluids partially dry the product.

The size of feed nozzle must be between 1 and 6 mm and the product must be less than 5 μm.

Energy consumption varies, according to the situation, between 30 and 500 kWh.tonne^{-1}.

The flow rate could range between a few kilograms per hour to 10 tonnes per hour according to the product, the desired fineness and the type of the equipment.

The gaseous flow ranges between 200 and 15,000 m^3.h^{-1} in operating conditions.

Typically, in equipment using fluid energy, the fluid drives particles that break by mutual collision. Movement can be spiral and could also oppose two fast jets or even direct the jet onto a target.

The fluid used is normally a gas.

The gas could be compressed air or even superheated vapor that pulls particles at a speed that can reach 300 m.s^{-1}. Particles break because they inter-collide or because they hit a target.

We then obtain a product whose size has a magnitude of micrometers.

In the pepper mill, grains are carried by an air current along the rotor's axle (horizontal). They hit radial elements at the edge of the rotor and, as such, break.

Tanaka [TAN 72], for gas jet equipment, suggests a theoretical description that leads to an extrapolation law relating the equipment's capacity to its energy consumption. An anonymous author suggests [ANO 83] a description of various types of compressed air mills.

4.8. Dispersers (in a liquid)

4.8.1. Emulsion troughs

Dispersion of one liquid in another could occur through constriction, where speed gradients (pronounced gradients) are very high and break the drops in suspension, to give droplets that are a micron in size (emulsion).

Similarly, opposite disc equipment has a feed along the disks' axle (vertical) that turns at different speeds (but in the same direction) or even where one does not turn. The centrifugal force takes the product to the periphery. The interval between the two discs sets the fineness of the product obtained. This equipment is suitable for organic products that are not hard but are resistant (maize groats, wood pulp).

This equipment also serves to disperse a powder in a liquid.

NOTE.–

The same principle of opposite discs is used for milling and dispersal of solids in a liquid.

We feed the interval found between two discs that are opposite and that are driven by a relative speed in relation to each other, but which turn in the same direction. The feed is located at the center of the discs.

The interval between discs is low but finite. We cannot reduce it indefinitely but when grains become too small, they are no longer subjected to shearing.

In mining industry laboratories, we use metallic discs fed in the center and whose axle is horizontal, used solely for wet milling. Cylinder diameters can reach 900 mm. We could hope for a fineness having a magnitude of 40 μm.

This equipment is suitable for soft products.

4.8.2. *Paint grinders–dispersers*

The paint mill is made up of three rollers turning faster and faster in the same direction of the moving product. Different speeds have a dual purpose:

– to bring about sharing between cylinders;

– to ensure that the product's progress is from the slow roller to the faster roller.

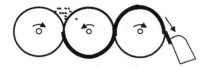

Figure 4.9. *Paint mill*

Choice in Comminution Equipment: Mechanical Plant Processing

5.1. Characteristics of divided solids

5.1.1. *Particle size distribution*

Let x_i be the sieve opening with index i.

By numbering the sieves from top to bottom in decreasing order by size:

$$x_{i-1} > x_i$$

Let m_i be the retained mass between the sieves i–1 and i. For Gaudin [GAU 26], log m_i is a function of log x.

In what follows, we will refer to the biggest grains as *granules* and will call everything else *fines*.

We use, for fines:

$$\log m_i = k \log x_i + \log C \quad \text{or} \quad m_i = C x^k$$

If we make sure that $\sum_{1}^{n} m_i = 1$, then m_i is the *mass* retained on the sieve with index i between sieves of magnitudes i and i–1. Note that m_i values measure the *frequency* of the size distribution (S.D.).

The constant k is the slope of the rectilinear part of the curve. It is the distribution *module* for particle diameters. This distribution is a frequency distribution here.

The higher the distribution rich in ends, the lower the module k.

The measurement in granulometry is made:

– in the sieves between 100 μm and 100 mm;

– in an optical microscope between 0.5 μm and 1 mm;

– by classifying the settling in a fluid between 5 μm and 0.5 mm;

– there are automatic equipments that give the granulometry of a bulk solid by using optical properties of its suspension in water.

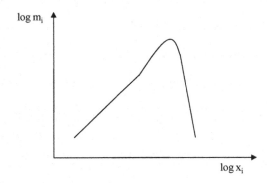

Figure 5.1. *Frequency distribution of particle diameters*

5.1.2. *Solid's suitability for grinding (grindability)*

This property (see [CHA 65]:

– increases with the product's fragility;

– decreases with the product's Mohs hardness.

1) Fragility:

The product does not bear the slightest strain without breaking (brittle product). Deformations imposed in mills are very quick, about 0.1 s. Stress required does not have time to spread and be absorbed in the particle's mass and remains concentrated in the restricted domain where it is intense. A brittle product cannot resist this. In general, ores have brittle behavior and do not respond to stress with plastic deformation.

2) Mohs hardness:

A solid can scratch another solid that is less hard (see Appendix 1).

A hard product in the Mohs sense is generally difficult to grind. Ores with a hardness that is less than or equal to 5 are considered soft, for example, bauxite, clay, magnesite, phosphates, iron oxide pigments, sulfur. Ores with a hardness greater than 6 are abrasive for grinding surfaces. They are considered to be hard when they come to grinding.

NOTE.–

The opposite of brittleness is plasticity like modeling clay or beeswax. The way to transform plasticity into brittleness is to cool the product, for example, by cooling it with ice or in extreme cases by plunging it in bubbling liquid nitrogen.

NOTE.–

In terms of energy used, the suitability in grinding of a bulk solid is quantified by its Bond Index. This index (which is energy per unit mass of bulk solid) *is lower as the solid is more easily ground.*

5.2. Machine selection

5.2.1. *General points about machine selection*

All comminution equipment function correctly if the product is dry. When a product is moist, it can become sticky and limit the capacity according to each device.

Hardness using Mohs scale (see Appendix 1) is directly linked to abrasiveness. We can without any doubt show the practical qualification of this scale by using the following qualifiers given in Table 5.1.

Hardness qualification	Mohs index I_M
Soft	$0 < I_M < 4$
Moderate	$4 \leq I_M < 7$
Hard	7–10

Table 5.1. *Hardness qualification*

The connection between hardness and abrasiveness is the fact that a solid A that scratches another solid B has an index where $I_{MA} > I_{MB}$. It is the method that was used to organize different solids using the Mohs scale.

Crushers operate generally by compression. However, some crushers operate by collision (hammer crusher).

Grindstone mills often lead to compression grindstones (track mills with whose axle is), while certain grindstones lead to shearing.

Rod and ball mills operate by collision and by shearing.

Autogenous mills encapsulate three types of action: collision, compression and shearing.

Attrition mills (vibratory or non-vibratory) operate as their name indicates and, in addition, by collision, just like micronizer.

5.2.2. *A criterion in machine selection: solid's hardness*

For crushing pieces or grinding hard products (Mohs hardness greater than 7), compression between two solid surfaces is used or even shearing between two solid surfaces sliding one against the other.

For soft products (Mohs hardness less than 4), we use collision on a solid surface. This is the situation for hammer, pin, ball or rod mills.

For plants or plastics, slicing with fixed knives on a revolving shaft is used.

5.2.3. *Machine selection according to flow rate (and hardness)*

1) Crushing:

– Hard products:

For flowrate of up to 1,000 ton.h^{-1} and more, gyratory and jaw crushers are suitable and are used for many ores.

– Soft products:

For flows of up to 300 ton.h^{-1}, the roller crusher is used. For superior flows (up to 1,500 ton.h^{-1}), the hammer crusher is recommended (limestone, bauxite, brick, cork). Cork is processed in a hammer device, because it is too elastic for a device with rollers.

2) Grinding:

– Hard products:

For flows of 20–500 ton.h^{-1}, the ball mill is suitable (silica, feldspar, refractories, cement, coal). For economic operation, the size of product obtained must not be less than 100 μm.

– Soft products:

For up to 10 ton.h^{-1}, the hammer mill achieves a product size that is never less than 10 μm (chalk, talcum, pigments, dyes, pepper, sugar, resins, press cakes).

From 10 to 50 ton.h^{-1}, the track mill carries out, with grinding bodies rolling, real kneading of the product. The size obtained is in millimeters and, at best, 100 μm (bauxite, gypsum, graphite, starch, phosphate, talcum, sulfur, inorganic salts).

From 50 to 300 ton.h^{-1}, the roller mill is suitable and it is used particularly for agro-food products (which are especially soft).

3) Autogenous mill:

It replaces the crusher-mill combination and delivers a product whose average size is not less than 300 or 400 μm.

4) Mills for ultrafines:

Flow in some of this type of equipment could reach 10 ton.h^{-1}.

– Hard products:

Gas jet equipment processes alumina or mica.

– Soft products:

The attrition mill and liquid shearing equipment is suitable for pigments and dyes, especially for the paint industry.

CONCLUSION.–

Table 5.2 extracted from Kelleher [KEL 59], and improved by Formanek [FOR 78], summarizes the requirements for machine selection.

In Gaudin [GAU 26], we find indications about the shape of particles obtained according to the equipment used.

Type	Maximum Capacity (ton.h⁻¹)	Diameter of maximum allowable piece (mm)	Possible Reduction Ratio	Hardness	Adhesiveness
Jaw crushers	1,000	1,350–1,500	4–10	Moderate and hard	Slightly sticky
Gyratory crusher	4,000	1,350	6–20	Moderate and hard	Non-sticky
Crusher with smooth cylinders with identical speeds	300	300	2–4	Moderate and hard	Sticky
Crusher with smooth cylinders with different speeds	100	90	4–10	Friable and moderate	Very sticky
Toothed roller crusher	500	1,000	4–5	Friable and moderate	Very sticky
Impact crusher	1,360	500	15–50	Soft and moderate	Sticky
Hammer crusher	1,500	1,300	10–30	Soft and moderate	Sticky
Grindstone mills	5–15	30	100	Soft to hard	Dry or as a pulp
Autogenous or tumbling mill	350	600	1,000–2,500	Moderate to hard	Very sticky
Rod mill	375	30	10–100	Moderate to hard	–
Ball mill	500	35	5–300	Soft to hard	–
Vibrating ball mill	6	25	500	Soft to hard	–
Attrition crusher	0.1	25	100	Soft to hard	–
Pulverizer	5	0.050	10–20	Soft to moderate	Slightly sticky

Table 5.2. *Characteristics of mills and crushers*

5.2.4. *Precautions to take according to the nature of the processed solid*

The following situations could arise:

1) The bulk solid is completely dry and does not have many fines. It is the simplest situation and the product to be processed simply passes through the device by coming out with a size distribution such that, for example, 80% by mass has a diameter less than the specification d_s.

2) The divided solid is almost dry but the slight moisture that it does have makes it sticky and it adheres to grinding surfaces. It must be watered during milling (and especially crushing).

3) A wet solid can be dried by having it circulate in the mill:

– superheated vapor;

– hot air.

4) The fed solid was dispersed beforehand in a liquid (often water) after, for example, physical or chemical processing. This dispersion is called a pulp or a slurry. This type of milling is known as wet milling as opposed to dry milling described in (1).

5) The solid is thermo-sensitive. This could lead to cooling as all milling is accompanied by heat emission. The coolant could be:

– dry, cold air;

– in wet operation, the presence of water is enough to absorb the heat released and, therefore, to control overheating.

6) A soft solid could not be milled properly in this state. Cryogenic cooling will make it breakable, brittle and therefore easily milled.

7) The moderate presence of fines, in humid way, increases the capacity by giving the pulp viscosity, which prevents particles that are trapped between two balls from escaping. On the other hand, if the pulp is too viscous, it acts as a cushion between balls and reduces the milling efficiency. As a result, dilution must be increased if the amount of fines increases.

5.3. Processing plants – the plant cell

5.3.1. *Structure of a plant cell*

By inspecting a plant cell, we find:

1) An amorphous pectin "cement" that maintains cohesion between cells. This layer is a cellular secretion that is made of protopectin, which is a copolymer of galacturonic acid and its methyl ester. Pectic cement could contain lignin in a proportion that increases with the plant's age. It results that, wood could contain up to 40% lignin. Contrary to protopectin which depolymerizes with heat, lignin is temperature insensitive.

2) A cellulose envelope that is equally insensitive to heat. It is a cellular secretion.

3) The water-permeable cytoplasmic membrane.

4) The cytoplasm which is also called the cell body. Cytoplasm includes transparent hyaloplasm that is mainly composed of water. Inside the hyaloplasm, there are inclusions which are basically vacuoles and plastids.

5) Vacuoles, plentiful in the flesh of fruits and rare in seeds, are pockets of aqueous solutions of sugars, acids and salts. This is where we find beet sugar and cane sugar. The vacuole membrane is semi-permeable. For water, osmosis laws apply. For solutes, crossing the membrane is done by a preliminary combining with protein carriers called permeases. In other words, the simple chemical potential, gradient does not express translate into the transfer of solutes but only the transfer of water. Vacuoles give the cell turgor phenomenon that we will examine further.

6) Plastids and inert enclaves. These enclaves could contain:

– starch (amyloplastids), found in seeds;

– proteins: hard wheat (rich in proteins) is used for making pasta and soft wheat (rich in starch) used to make bread;

– lipid droplets (in oilseeds);

– pigments (chloroplasts contain chlorophyll);

– inorganic corpuscles;

– air (especially in some plants growing in softwater like milfoil, or even in some oilseeds).

7) Cytoplasm, furthermore, contains:

– the cell nucleus;

– various inclusions that we will not mention.

8) Parenchyma: It is the connective tissue that forms the 'flesh' in fruit, roots and stems.

9) Extracellular air: There could be air pockets in the parenchyma between cells. As such, in an apple, the volume taken up by air could reach 20–25% of the fruit's volume. In peaches, air volume gets to 15% and only 1% in potatoes.

5.3.2. *Effect on the taste of fruit*

Mechanical resistance in the cellulose layer of the cellular membrane defines the nature of fracture surfaces of the parenchyma after deformation and fracture of the pectic cement.

If the cellulose layer is only slightly resistant, the cellular membrane breaks and liberates the cell contents and, particularly, the sweet juices of its vacuoles. This is the reason why some fruits taste sweet when chewed.

If, inversely, the cellulose layer is resistant, the cell will remain intact during fracture because cracks in the pectic cement will spread among themselves. The fruit will then have a mealy taste.

5.3.3. *Effect on vegetable oil extraction*

Air inclusions in oilseed cells are obstacles for oil or miscella migration. A thick cell membrane also makes oil extraction difficult.

If cells are small in size, the flow of oil or miscella between them will be much more difficult because the permeability of a porous medium is inversely proportional to the grain diameter squared.

Table 5.3 organizes seeds by their increasing difficulty for its oil extraction (by solvent or by pressing).

Characteristics of seeds	Amount of air	Thick membrane	Small cells
Soya	no	no	no
Sunflower	no	no	no
Castor	no	no	no
Rapeseed	yes	no	yes
Coriander	yes	yes	yes

Table 5.3. *Oil extraction in seeds*

5.3.4. *Turgescence pressure*

Let us write, with the help of chemical potentials, the liquid–vapor equilibrium of pure water:

$$\mu_V\left(T, \pi(T)\right) = \mu_L\left(T, \pi(T)\right)$$

π is the tension of pure water vapor at a temperature T.

More specifically, this relationship can be written as:

$$\mu_V^*\left(T, P_o\right) + RTLn\left[\frac{\pi}{P_o}\right] = \mu_L^*\left(T, P_o\right) + V_e\left(\pi - P_o\right) \qquad [5.1]$$

P_o is a reference pressure and V_e is the molar volume of water.

Let us now write the equilibrium between the surrounding atmosphere and the liquid in the vacuole, where water has a molar fraction x, and pressure in the vacuole being P_L:

$$\mu_V\left(T, P_V\right) = \mu_L\left(T, P_L, x\right)$$

P_V is the partial water vapor pressure in the surroundings.

More specifically:

$$\mu_V^* \left(T, P_o \right) + RT \, Ln \left[\frac{P_V}{P_o} \right] = \mu_L^* \left(T, P_o \right) + V_e \left(P_L - P_o \right) + RT \, Ln \, a_e \qquad [5.2]$$

a_e is water's activity in the vacuole.

Subtracting component by component in equation [5.1] from equation [5.2]:

$$RT \, Ln \left[\frac{P_V}{\pi} \right] = RT \, Ln \, a_e + V_e \left(P_L - \pi \right)$$

P_V/π is the relative humidity ε of the gas phase.

$RT \left[Ln \, a_e \right] / V_e$ is the osmotic pressure π_0 of the solution.

From:

$$P_L - \pi = \pi_o + \frac{RT}{V_e} Ln \, \varepsilon$$

Let us set:

$$P_L - \pi = P_T$$

P_T is the turgescence pressure.

$$P_T = \pi_o + \frac{RT}{V_e} Ln \varepsilon = \frac{RT}{V_e} Ln \frac{\varepsilon}{a_e}$$

Practically, the turgescence pressure could reach 15 bars. To measure this pressure experimentally, we use a hygrometer that gives the relative humidity ε of the surroundings, kills the cells by freezing them and then measure water activity a_e in the exudate (for example, by using calibrated cell paper).

As such, a plant placed in a dry atmosphere will wither because its cell vacuoles empty their water as the turgescence pressure is low.

If the turgescence of the plant is low, it is soft and easily deformed like a flaccid rubber ball. Its resistance to compression or traction is low. Applying a force will lead to a fracture after a significant amount of deformation.

5.3.5. *Action of heat alone on plant cells*

An increase in temperature to 90°C:

– coagulates protein inclusions and precipitates phosphatides; these products are no longer carried in the juice or in oil, which improves quality;

– kills bacteria as well as hydrolyzing enzymes that generate fatty acids (that gives oil a bad taste).

If a higher temperature is maintained during extraction, the viscosity of juices, oils or miscella will be reduced and extraction is easier.

Temperature alone has no effect on cellulose and lignin.

5.3.6. *Action of moisture alone on plant cells*

Mutual cohesion of cellulose chains (which is a polysaccharide) of the cellular membrane is due to transverse hydrogen bonds involving OH groups from the cellulosic molecule. However, water establishes the same type of bonds with its OH groups to the detriment of bonds between chains.

The outcome is that the cellulosic membrane swells. Basically, it is well understood that wood swells with humidity and that cotton (which is pure cellulose) easily fills with water. The uncoupled cellulose chains can allow for non-polar organic liquids to pass, that is liquids without any affinity for cellulose. The cellulosic membrane will act as a porous membrane in a way that:

– moderate humidity will open "pores";

– excessive humidity will fill pores and prevent hydrophobic liquids (solvents, oil) to pass through the membrane similar to how a blotting paper that is filled with water is oil impermeable.

This is why hydrothermal processing of oilseeds is followed by drying before oil extraction.

5.3.7. *Simultaneous actions of humidity and heat*

Hot humidification depolymerizes protopectins of the cellular membrane that become transformed into hydro-soluble pectins, found dissolved in aqueous juices or as sludge in oils.

The peptic cement being softened, the plant tissue acquires plasticity and this explains the possibility to flatten oilseed flours to microscale. It is also the reason why cooked vegetables and fruits are much softer than in their raw state.

5.3.8. *Effects of freezing and dehydration on plant cells*

These two operations, separately or together, kill cells. The selective permeability of membranes disappears and penetration by foreign substances becomes possible, as well as making the exudation of internal substances easier.

5.4. Mechanical preparation of plants

5.4.1. *Cleaning and preliminary scrubbing*

Rocks, earth and sand in plants and seeds lead to untimely material wear and their presence is detrimental to the quality of the final product.

Plant debris (bracts, leaves, twigs from mechanical gathering) decompose during storage because, if there is an excess of moisture in seeds, there is fermentation and overheating. Storage must then be ventilated (wheat, cotton).

Metallic debris could lead to sparks that may generate fires and even an explosion.

Preliminary scrubbing is a form of separation that can be done:

– by a pneumatic path;

– by screening;

– by a hydraulic path (washing);

– by a magnetic path.

1) Pneumatic scrubbing:

Green peas being poured into an ascending air current will have any dust carried away.

Air currents carry away light hulls of oilseeds, plant debris and, of course, dust. It is possible to separate debris from hulls and kernels by placing the product in a drum swept by air and wit turning bats.

2) Screening:

Almonds are denser than hulls and pass through the plane of the screen if the screen has a reasonably sized opening.

Stone debris are denser than almonds and could equally be separated by screening.

3) Washing:

Washing is not suitable for fruits and vegetables and is avoided for seeds.

Spinach and asparagus are immersed in a turbulent water current that removes earth and sand. Water may be chlorinated.

In a gutter, water pulls out peas and leaves the earth and sand behind.

Light plant debris float in a tank full of water.

Plants could be rinsed under a water jet in a rotating drum.

Citrus fruits are brushed during washing.

A detergent solution bath removes pesticides from apples.

4) Magnetic scrubbing:

A magnet placed in a different direction attracts ferrous-based debris. There are two types of magnets:

– the turning drum with continuous debris separation;

– the plate that is cleaned manually.

5.4.2. *Grinding and slicing (fruit, seeds, tubers)*

We can distinguish:

– end grinding of a fruit;

– rolling oilseed almonds;

– cutting sugar beet chips.

1) Grinding:

This operation can take place using a hammer or a disc mill. The capacity of a hammer mill is much higher but it does not grind as fine as a disc mill does.

We can then obtain the nectars of certain fruits directly. With more general milling of entire apples, hard debris remain which ultimately make the pressing operation which produces juice, easier because they avoid clogging the product bed.

Coffee is ground in mills with grooved cylinders and the grain size achieved is approximately between 0.4 and 0.8 mm.

2) Rolling:

Diffusion theory in a thin plate shows that the Fourier number increase is inversely proportional to the plate thickness squared. This means that, during solvent extraction, depletion of the product will be made easier if it is shaped into thin plates.

Filtration theory shows that the pressure required for a liquid to pass through a plate increases proportionally to the plate thickness. Pressing theory makes use of this fact.

For these reasons, there is every advantage to processing particles of product into thin plates, but a third reason is that the porosity of a bed of loose plates is much greater than the porosity of a bed of spheres. The interstitial spaces that are then maintained make it easier for the liquid phase to circulate as much during pressing as in solvent extraction.

This is why oilseed flours are subjected to laminating in a smooth roller mill. Tiny flakes escape whose thickness is a fraction of a millimeter.

The two cylinders turn at different speeds and the faster cylinder pulls the product to the surface. The thickness of oilseed flour flakes or even the fineness of cereal flour is adjusted by acting on the gap separating the surfaces of the two cylinders.

Cereals which are slightly moist give little mutual cohesion when milling flour grains, while oilseeds, with preliminary moisture added, are transformed into micro-flakes that are relatively coherent and that do not break down into independent particles. Oil, still enclosed on the inside of cells, but having been partially exuded under cylinder pressure, can help with flake cohesion.

Oilseed flour production using smooth cylinder mills is about 0.6 ton.day^{-1} per centimeter of cylinder length. As such, cylinders turning at 300 rev.mn^{-1} and a diameter of 60–80 cm will process between 200 and 250 ton.day^{-1} of flour and the mill will consume between 2.5 and 3.5 kW per ton.h^{-1} of flour.

3) Cutting:

For the same reasons as oilseed flour, sugar beets are fragmented into chips (cosettes) before sugar extraction. Chips are shaped like tiles with V-shaped sections, the height of the V ranges between 2 and 6 mm and the thickness of these "tiles" could be adjusted to between 0.1 and 2.3 mm; this operation is carried out in devices equipped with knives. Sugar in the chips is then extracted at $76°C$ by a solvent that is none other than water.

5.4.3. *Hulling seeds*

All fruits have at least one seed made up of an almond and enclosed in a hull. The hulling operation is made up of separating the almond and the hull and it is important in the oilseed industry.

The most popular equipment used for this operates under three distinct principles:

– collision that bursts the seed;

– combined squeezing (or not) with shearing;

– abrasion.

1) Cutting equipment:

A recent type is made up of a drum with knives turning on the inside of a fixed cylinder (the concave) also holding knives on a fraction of its perimeter. The gap between the drum and concave is adjustable, which helps to control the fineness of fragments obtained. This equipment is used for cotton.

Another type of equipment (used for sunflower) is made up of a disc with a vertical axle with radially fixed blades fixed and inclined in relation to the disc's surface. The product is fed from the top, subjected to blade action and is projected to the periphery by centrifugal force. It is collected in the bowl that holds the disc. Rotation of the disc axle is also the bowl's axis of symmetry (immobile). The product flows by gravity along the wall of the bowl and it is collected at the tip of the cone.

Equipment with knives are collision devices.

2) Squirrel Cage Husker:

A collision device that can be used for sunflowers. The rotor is a squirrel cage that projects seeds against a cylindrical stator equipped with longitudinal channels on one part of its perimeter.

3) Centrifuge Husker:

This system is used for rapeseeds and it is called a desheller. Seeds are placed in the center of the rotor, ejected along the grooves of the latter and projected against the cylindrical target surrounding the rotor. Here, the seed undergoes only one collision. The rotor rotates at 3,000 rev.mn^{-1}.

4) Disc Husker:

This equipment operates by squeezing and shearing. Seeds are placed in the center of the two discs that are facing each other. The two discs turn at different speeds and they are streaked radially. Seeds are squeezed and ejected to the periphery by a centrifugal force. The face of each disc is slightly conical to allow for placing the product in the center. The section that allows seeds to pass retracts in the direction of the periphery. Disc diameter ranges from 0.5 to 1 m.

This equipment is suitable for cotton and sunflowers.

5) Roller Husker:

The seed is broken down between two grooved rollers that squeeze and shear seeds because they do not turn at the same speed. Soya can be processed in this way.

6) Abrasion Husker:

It is used for chestnuts.

5.4.4. *Peeling*

Potatoes, which have a spheroidal shape and are prone to collisions, are peeled in a turning chamber with abrasive walls, just before their usage.

Yellow peaches, carrots, turnips and salsify are peeled by immersing them in a hot alkaline solution (95°C). These plants are then rinsed under a high-pressure water jet.

Onions are peeled by twisting, that is using a very hot air current.

Green peas are extracted by threshing them in a drum.

5.4.5. *Plant stem fractionation (sugar cane)*

Sugar cane fractionation requires two steps:

– passing through the cane-cutter;

– reduction and disintegration in the hammer mill.

Canes slide on a metallic tablet inclined roughly 30° horizontally on the top of the cane bed. A perpendicular horizontal axle, turns in the same direction as the canes. This axle has six knives on a circle, with 5 cm spaces between the circles in the first device and a 2.2 cm space in the next device. The shaft turns at 500 rev.mn^{-1}.

The cane-cutter can be substituted with a defibrer made up of two cylinders equipped with chevron teeth.

In both cases, the power consumed is approximately 20–25 kW per ton.h^{-1} of cane.

Cane then goes into hammer mills that tear the tissues. The rotor, on a horizontal axle, turns at a speed of approximately 1,000 rev.mn^{-1} and is made of pivoting hammers. In the lower force of its periphery, the rotor is facing an anvil, made of parallel horizontal bars, that partially surrounds the rotor and its hammers. The space between the hammer's trajectory and the anvil's bars drops progressively from 25 to 5 mm starting from when the product is introduced at the inlet until the outlet.

The power consumed is, even here, from 20 to 25 kW per ton.h^{-1} of cane. Equipment and current power is 1 MW (1 MW = 1,000 kW).

The appearing density of the prepared bagasse depends on the production location and varies from 200 to 350 kg.m^{-3}, while the appearing density of loose cane was only ranging from 120 to 550 kg.m^{-3}.

5.4.6. *Fruit fractionation (threshing)*

Fruit generally contains a nut (kernel) surrounded by pulp.

Separating the pulp from the kernel takes place in the depulping machine (also called dekerneling machine in the olive oil industry). This equipment is made up of a cylinder with perforated sheet metal or even a type of squirrel cage with bars. In the inside of the cylinder or the cage turns a shaft covered with pallets or bats.

The ferrule (cylinder or cage) turns to generate centrifugal acceleration. The shaft with pallets also turns. The difference in respective rotation speeds is marked, in such a way that the product is subjected to significant and repeated collisions. The shaft with pallets could even turn in the opposite direction to the ferrule. The slurry that contains the pulp goes through the ferrule and kernels are removed at the opposite end of the feed. The tomato juicer operates on the same principle.

The threshing principle was just described and also used to extract green peas separating hull debris sticking to almonds and, finally, to separate grains of wheat from the husk and the straw.

The stoning machine acts as a preliminary operation in making jams (peach, apricot, tomato) or nectars.

Even though no fruit is involved, dried tobacco leaves are subjected to a threshing operation to separate ribs from the parenchyma.

APPENDICES

Appendix 1

Mohs Scale

Nature of bulk solid	Mohs index
Wax	0.02
Graphite	0.5–1
Talcum	1
Diatomaceous earth	1–1.5
Asphalt	1.5
Lead	1.5
Gypsum	2
Human nail	2
Organic crystals	2
Lye flakes	2
Slaked lime	2–3
Sulfur	2
Salt	2
Tin	2
Zinc	2
Anthracite	2.2
Silver	2.5
Borax	2.5
Kaolin	2.5
Litharge	2.5
Sodium bicarbonate	2.5
Copper (coins)	2.5

Slaked lime	2–3
Aluminum	2–3
Quicklime	2–4
Calcite	3
Bauxite	3
Mica	3
Plastic materials	3
Barite	3.3
Brass	3–4
Limestone	3–4
Dolomite	3.5–4
Siderite	3.5–4
Sphalerite	3.5–4
Chalcopyrite	3.5–4
Fluorite	4
Pyrrhotite	4
Iron	4–5
Zinc oxide	4.5
Glass	4.5–6.5
Apatite	5
Carbon black	5
Asbestos	5
Steel	5–8.5
Chromite	5.5
Magnetite	6
Orthoclase	6
Clinker	6
Iron oxide	6
Feldspar	6
Pumice	6
Magnesia (MgO)	5–6.5
Pyrite	6.5
Titanium oxide	6.5
Quartz	7

Sand	7
Zirconium oxide	7
Beryl	7
Topaz	8
Emery	7–9
Garnet	8.2
Sapphire	9
Corundum	9
Tungsten carbide	9.2
Alumina	9.25
Tantalum carbide	9.3
Titanium carbide	9.4
Silicon carbide	9.4
Boron carbide	9.5
Diamond	10

We can then classify materials according to their hardness:

– Soft   1–3

– Fairly Soft   4–6

– Hard   7–10

Real Density of Loose Bulk Solids (kg.m^{-3})

A2.1. Plant products

Nature of product	Grains or seeds		Flour	
Flax	720		430	
Corn	720		640	
Cotton	530		400	
Soya	700	(chipped)	540	
Coffee	670	(green)	450	(roasted, ground)
Wheat	790			
Barley	620			
Rye	720			
Rice	800			
Oats	410			
Clove	770			

A2.2. Natural inorganic products

Nature of product	Granules		Powders (fine grinding)	
Bauxite	1,360	(run-of-mine)	1,090	
Gypsum	1,270		900	
Kaolin	1,024	(crushed)	350	(<10 µm)
Lead silicate	3,700		2,950	
Quicklime	850		430	
Limestone	1,570		1,360	
Phosphate	960		800	
Wood waste	350	(shavings)	320	(sawdust)

Sulfur	1,220		800	
Iron	4,950	(beads)	2,370	(filings)
Slate shale	1,390		1,310	
Sodium carbonate	1,060		480	
Coke	490		430	

A2.3. Manufactured products

Powdered sodium bicarbonate	690
Borax	1,700
Catalyst (cracking fluidized oil)	510
Ashes	700
Charcoal (grains)	420
Charcoal (crude)	900
Charcoal (graded)	700–800
Cement (clinker)	1,400
Cement (Portland)	1,520
Chipped copra shavings	510
Chipped copra shavings pressure screw outlet	465
Powdered dolomite	730
Soap flakes	160
Ground feldspar	1,600
Gravel	1,500
Dairy	2,000
Ground mica	210
Ground bone	1,200
Phthalic anhydride flakes	670
Glass beads	1,400
Iron oxide pigment	400
Zinc oxide pigment	320
Lead shot	6,560
Potatoes	700
Rubber scraps	370
Sand	1,350–1,500
Salt	1,200
Crystallized sugar	830
Crystallized copper sulfate	1,200
Powdered superphosphate	810

Bibliography

[AUS 71a] AUSTIN L.G., "Introduction to the mathematical description of grinding as a rate process", *Powder Technology*, vol. 5, p. 1, 1971.

[AUS 71b] AUSTIN L.G., LUCKIE P.T., ATEYA B.G., "Residence time distributions in mills", *Cement and Concrete Research*, vol. 1, pp. 241–256, 1971.

[AUS 72a] AUSTIN L.G., BHATIA V.K., "Experimental methods for grinding studies in laboratory mills", *Powder Technology*, vol. 5, p. 261, 1972.

[AUS 72b] AUSTIN L.G., LUCKIE P.T., "Methods for determination of breakage distribution parameters", *Powder Technology*, vol. 5, p. 215, 1972.

[AUS 72c] AUSTIN L.G., LUCKIE P.T., "The estimation of non-normalized breakage distribution parameters from batch grinding", *Powder Technology*, vol. 5, p. 267, 1972.

[AUS 72d] AUSTIN L.G., LUCKIE P.T., "Grinding equations and the Bond work index", *Transactions of the Society of Mining Engineers*, vol. 252, pp. 259–266, 1972.

[AUS 73a] AUSTIN L.G., BHATIA V.K., "Note on conversion of discrete size interval values of breakage parameters S and B to point values and vice versa", *Powder Technology*, vol. 7, p. 107, 1973.

[AUS 73b] AUSTIN L.G., SHOJI K., EVERETT M.D., "An explanation of abnormal breakage of large particle sizes in laboratory mills", *Powder Technology*, vol. 7, p. 3, 1973.

[AUS 74] AUSTIN L.G., SHOJI K., BHATIA V.K. et al., "Extension of the empirical Alyadin equation for representing batch grinding data", *International Journal of Mineral Processing*, vol. 1, p. 107, 1974.

[AUS 75] AUSTIN L.G., LUCKIE P.T., WIGHTMAN D., "Steady-state simulation of a cement-milling circuit", *International Journal of Mineral Processing*, vol. 2, pp. 127–150, 1975.

[AUS 76] AUSTIN L.G., SHOJI K., LUCKIE P.T., "The effect of ball size on mill performance", *Powder Technology*, vol. 14, p. 71, 1976.

[AUS 79] AUSTIN L.G., JINDAL V.K., GOTSIS C., "A model for continuous grinding in laboratory hammer mill", *Powder Technology*, vol. 22, pp. 199–204, 1979.

[AUS 80] AUSTIN L.G., VAN ORDEN D.R., PÉREZ J.W., "A preliminary analysis of smooth roll crushers", *International Journal of Mineral Processing*, vol. 6, p. 321, 1980.

[AUS 81a] AUSTIN L.G., BAGGA P., "An analysis of fine dry grinding in ball mills", *Powder Technology*, vol. 28, p. 83, 1981.

[AUS 81b] AUSTIN L.G., SHAH J., WANG J. et al., "An analysis of ball-and-race milling. Part I. The hardgrove mill", *Powder Technology*, vol. 29, p. 263, 1981.

[AUS 81c] AUSTIN L.G., VAN ORDEN D.R., MC WILLIAMS B. et al., "Breakage parameters of some materials in smooth roll crushers", *Powder Technology*, vol. 28, p. 245, 1981.

[AUS 82a] AUSTIN L.G., LUCKIE P.T., SHOJI K., "An analysis of ball-and-race milling. Part II. The Babcock E. 17 Mill", *Powder Technology*, vol. 33, p. 113, 1982.

[AUS 82b] AUSTIN L.G., LUCKIE P.T., SHOJI K., "An analysis of ball-and-race milling. Part III. Scale-up to industrial mills", *Powder Technology*, vol. 33, p. 127, 1982.

[AUS 83a] AUSTIN L.G., BRAME K., "A comparison of the Bond method for sizing wet tumbling ball mills with a size-mass balance simulation model", *Powder Technology*, vol. 34, p. 261, 1983.

[AUS 83b] AUSTIN L.G., ROGOVIN Z., ROGERS R.S.C. et al., "The axial mixing model applied to ball mills", *Powder Technology*, vol. 36, p. 119, 1983.

[AUS 84] AUSTIN L.G., LUCKIE P.T., SHOJI K. et al., "A simulation model of an air-swept ball mill grinding coal", *Powder Technology*, vol. 38, p. 255, 1984.

[BAR 71] BARBERY G., "Regression analysis method for estimating the parameters of the three-parameter size distribution equation", *Transactions of the Society of Mining Engineers of A.I.M.E.*, vol. 250, p. 187, 1971.

[BAR 92] BARBERY G., "Liberation 1, 2, 3: theoretical analysis of the effect of space dimension on mineral liberation by size reduction", *Mineral Engineering*, vol. 5, no. 2, p. 123, 1992.

[BEC 99] BECKER M., SCHWEDES J., "Comminution of ceramics in stirred media mills and wear of grinding beads", *Powder Technology*, vol. 105, p. 374, 1999.

[BEC 01] BECKER M., KWADE A., SCHWEDES J., "Stress intensity in stirred media mills and its effect on specific energy requirement", *International Journal of Mineral Processing*, vol. 61, p. 189, 2001.

[BER 61] BERGSTROM B.H., SOLLENBERGER C.L., MITCHELL W., "Energy aspects of single particle brushing", *Trans. ASME*, vol. 220, pp. 367–372, 1960.

[BON 43] BOND F.C., "Wear and size distribution of grinding balls", *Transactions of the American Institute of Mining and Metallurgical Engineers*, vol. 153, p. 373, 1943.

[BON 49] BOND F.C., "Standard grindability tests tabulated", *Transactions of the American Institute of Mining, Metallurgical and Petroleum Engineers Incorporated*, vol. 183, p. 313, 1949.

[BON 50] BOND F.C., WANG J.-T., "A new theory of comminution", *Transactions of the American Institute of Mining and Metallurgical Engineers*, vol. 187, p. 871, 1950.

[BON 52] BOND F.C., "The third theory of comminution", *Transactions of the American Institute of Mining, Metallurgical and Petroleum Engineers Incorporated*, vol. 193, p. 484, 1952.

[BON 54] BOND F.C., "Crushing and grinding calculations", *The Canadian Mining and Metallurgical Bulletin*, vol. 47, p. 466, 1954.

[BON 58] BOND F.C., "Grinding ball size selection", *Mining Engineering*, vol. 10, p. 591, 1958.

[BON 60] BOND F.C., "Confirmation of the third theory", *Transactions of the American Institute of Mining, Metallurgical and Petroleum Engineers Incorporated*, vol. 217, p. 139, 1960.

[BON 61a] BOND F.C., "Crushing and grinding calculations. Part I", *British Chemical Engineering*, vol. 6, no. 6, p. 378, 1961.

[BON 61b] BOND F.C., "Crushing and grinding calculations. Part II", *British Chemical Engineering*, vol. 6, no. 6, pp. 543–548, 1961.

[BRO 57] BROADBENT S.R., CALLCOTT T.G., "Coal breakage processes IV. An exploratory analysis of the cone mill in open-circuit grinding", *Journal of the Institute of Fuel*, vol. 30, p. 18, 1957.

[BUN 92] BUNGE F., PIETZSCH M., MÜLLER R. *et al.*, "Mechanical disruption of arthobacter S.P. DSM 3747 in stirred ball mills for the release of hydantoin-cleaning enzymes", *Chemical Engineering Science*, vol. 47, no. 1, p. 225, 1992.

[CAL 67] CALLCOTT T.G., "Solution of comminution circuits", *Transactions of the Institution of Mining and Metallurgy*, vol. 76, pp. C1–C11, 1967.

[CHA 56] CHARLES R.J., "High velocity impact in comminution", *Transactions of the American Institute of Mining and Metallurgical Engineers*, vol. 205, p. 1028, 1956.

[CHA 65] CHANDLER R.L., "Grindability tests for coal. Part I. The development of grindability tests", *The British Coal Utilisation Research Association Bulletin*, vol. 29, no. 10, p. 333, 1965.

[COG 34] COGHILL W.H., DE VANEY F.D., O'MEARA R.G., "Advantage of ball (rod) mills of larger diameters and advantage of improved bearings", *Transactions of the American Institute of Mining, Metallurgical and Petroleum Engineers Incorporated*, vol. 112, p. 79, 1934.

[DOR 44] DORRIS T.B., "How solid carbon dioxide assists in grinding low-melting waxy or plastic solids", *Chemical and Metallurgical Engineering*, vol. 51, p. 114, 1944.

[DUR 88] DURMAN R.W., "Progress in abrasion-resistant materials for use in comminution processes", *International Journal of Mineral Processing*, vol. 22, p. 381, 1988.

[DUR 16] DUROUDIER J.-P., *Liquid–Solid Separators*, ISTE Press Ltd, London and Elsevier Ltd, Oxford, 2016.

[ENG 87] ENGELS K., "Bead mill: a state of the art survey", *Paintindia*, p. 27, 1987.

[EPS 48] EPSTEIN B., "Logarithmico-normal distribution in breakage of solids", *Industrial and Engineering Chemistry*, vol. 40, p. 2289, 1948.

[FAI 53] FAIRS G.L., "A method of predicting the performance of commercial mills in the fine grinding of brittle materials", *Transactions of the Institution of Mining and Metallurgy*, vol. 63, p. 211, 1953.

[FOR 78] FORMANEK K., Techniques de broyage et consommation d'énergie, C.A.C.E.M. Session, 1978.

[FUE 85] FUERSTENAU D.W., VENKATARAMAN K.S., VELAMAKANNI B.V., "Effect of chemical additives on the dynamics of grinding media in wet ball mill grinding", *International Journal of Mineral Processing*, vol. 15, p. 251, 1985.

[GAO 93] GAO M.-W., FORSSBERG E., "A study on the effect of parameters in stirred ball milling", *International Journal of Mineral Processing*, vol. 37, p. 45, 1993.

[GAR 73] GARDNER R.P., SUKANJNAJTEE, "A combined tracer and back calculation method for determining particulate breakage functions in ball milling", *Part III. Powder Technology 7*, pp. 169–179, 1973.

[GAU 26] GAUDIN A.M., "An investigation of crushing phenomena", *Transactions of the American Institute of Mining and Metallurgical Engineers*, vol. 73, pp. 253–316, 1926.

[GAU 62] GAUDIN A.M., MELOY T.P., "Model and a comminution distribution equation for single fracture", *Transactions of the American Institute of Mining and Metallurgical Engineers*, vol. 223, p. 40, 1962.

[GÖL 87] GÖLL G., HANISCH J., "Zum Vergleich des Zerkleinerungsergebnisse bei der Druck-und Schlagbeanspruchung von Kornschichten", *Aufbereitrings Technik*, no. 10, p. 582, 1987.

[GÖT 56] GÖTTE A., ZIEGLER E., "Versuche zur Herabsetzung des Zerkleinerungs-Widerstands fester Stoffe durch gazförmige und dampförmige Zusatzmittel", *Zeitschrift des Vereines Deutscher Ingenieure*, vol. 98, p. 373, 1956.

[GOT 85] GOTSIS C., AUSTIN L.G., LUCKIE P.T. *et al.*, "Modeling of a grinding circuit with a swing-hammer mill and a twin-cone classifier", *Powder Technology*, vol. 42, p. 209, 1985.

[GRA 69] GRANDY G.A., GUMTZ G.D., HERBST J.A. *et al.*, "Computer techniques in the analysis of laboratory grinding tests", in WEISS A. (ed.), *A Decade of Digital Computing in the Mineral and Operation Research in the Mineral*, AIME, New York, 1969.

[GRI 20] GRIFFITH A.A., TAYLOR G.I., "The phenomena of rupture and flow in solids", *Philosophical Transactions of the Royal Society of London. Series A*, vol. 221A, pp. 163–198, 1920.

[GRU 83] GRUJIÉ M., OCEPEK D., SALATIÉ D., "Probleme der Optimierung in der Zerkleinerung", *Aufbereitungs-Technik*, no. 10, p. 575, 1983.

[GUP 81a] GUPTA V.K., HODOUIN D., BERUBE M.A. *et al.*, "The estimation of rate and breakage distribution parameters from batch grinding date for a complex pyritic ore using a back-calculation method", *Powder Technology*, vol. 28, p. 97, 1981.

[GUP 81b] GUPTA V.K., HODOUIN D., EVERELL M.D., "The influence of pulp composition and feed rate on hold-up weight and mean residence time of solids in grate-discharge ball mill grinding", *International Journal of Mineral Processing*, vol. 8, p. 345, 1981.

[HAD 38] HARDGROVE, *Trans. Am. Inst. Chem. Engrs*, vol. 34, p. 131, 1938.

[HAR 66] HARRIS C.C., "On the role of energy in comminution: a review of physical and mathematical principles", *Transactions of the Institute of Mining and Metallurgy Section C Mineral Processing and Extractive Metallurgy*, vol. 75, p. 37, 1966.

[HAR 69] HARRIS C.C., "A method for determining the parameters of the 3-parameter size distribution equation", *Transactions of the American Institute of Mining Metallurgical and Petroleum Engineers Incorporated*, vol. 244, p. 187, 1969.

[HEI 85] HEIM A., LESZCZYNIECKI R., AMANOWICZ K., "Determination of parameters for wet-grinding model in perl mills", *Powder Technology*, vol. 41, p. 173, 1985.

[HIX 91] HIXON L.M., "Select an effective size-reduction system", *Chemical Engineering Progress*, vol. 87, p. 36, 1991.

[HOG 75] HOGG R., SHOJI K., AUSTIN L.G., "Flow of particles through small continuous dry ball mills", *Transactions of the American Institute of Mining Metallurgical and Petroleum Engineers Incorporated*, vol. 258, p. 194, 1975.

[IWA 88] IWASAKI I., POZZO R.L., NATARAJAN K.A. *et al.*, "Nature of corrosive and abrasive wear in ball mill grinding", *International Journal of Mineral Processing*, vol. 22, p. 345, 1988.

[JIN 76] JINDEL V.K., AUSTIN L.G., "The kinetics of hammer milling of maize", *Powder Technology*, vol. 14, p. 35, 1976.

[JOH 65] JOHANSON J.R., "A rolling theory for granular solids", *Transactions of the ASME*, vol. 32, p. 842, 1965.

[JOO 96a] JOOST B., SCHWEDES J., "Zerkleinerung von Schmelzkorund und Mahlkörperverschleiß in Rührwerkskugelmühlen Teil 1", *Bericht D.K.G. C.F.I. Ceramic Forum International*, vol. 73, no. 6, p. 965, 1996.

[JOO 96b] JOOST B., SCHWEDES J., "Zerkleinerung von Schmelzkorund und Mahlkörperverschleiß in Rührwerkskugelmühlen Teil 2", *Bericht D.K.G. C.F.I. Ceramic Forum International*, vol. 73, nos. 7–8, p. 429, 1996.

[KAN 86] KANDA Y., SANO S., YASHIMA S., "A consideration of grinding limit based on fracture mechanics", *Powder Technology*, vol. 48, p. 263, 1986.

[KAW 88] KAWATRA S.K., EISLE T.C., "Rheological effects in grinding circuits", *International Journal of Mineral Processing*, vol. 22, p. 251, 1988.

[KEL 59] KELLEHER J., "Comminution theory and practice", *British Chemical Engineering*, vol. 4, p. 467, 1959.

[KEL 67/68] KELSALL D.F., REID K.J., RESTARICK C.J., "Continuous grinding in a small wet ball mill. Part. I. A study of the influence of ball diameter", *Powder Technology*, vol. 1, p. 291, 1967/1968.

[KEL 86] KELLERWESSEL H., "Betriebsergebnisse von Hochdruck-Rollenpressen", *Aufbereitungs-Technik*, p. 555, no. 10, 1986.

[KER 71] KERL J.F., "Autogenous grinding in laboratory tumbling mills", *Transactions of the Society of Mining Engineers of A.I.M.E.*, vol. 250, p. 188, 1971.

[KIC 85] KICK F., *Das Gesetz der proportionalen Wiederstande und seine Anwendung*, Leipzig, 1885.

[KIN 79] KING R.P., "A model for the quantitative estimation of mineral liberation by grinding", *International Journal of Mineral Processing*, vol. 6, p. 207, 1979.

[KLI 65] KLIMPEL R.R., AUSTIN L.G., "The statistical theory of primary breakage distributions for brittle materials", *Transactions of the Metallurgical Society of A.I.M.E.*, vol. 232, p. 88, 1965.

[KLI 70] KLIMPEL R.R., AUSTIN L.G., "Determination of selection-for-breakage fonctions in the batch grinding equation by nonlinear optimization", *Industrial and Engineering Chemistry Fundamentals*, vol. 9, p. 230, 1970.

[KLI 77] KLIMPEL R.R., AUSTIN L.G., "The back-calculations of specific rates of breakage and non-normalized breakage distribution parameters from batch grinding date", *International Journal of Mineral Processing*, vol. 4, p. 7, 1977.

[KLI 84] KLIMPEL R.R., AUSTIN L.G., "The back-calculation of specific rates of breakage from continuous mill data", *Powder Technology*, vol. 38, p. 77, 1984.

[KOL 93] KOLB G., "Fine grinding of ceramic glazes in agitator beads mills", *C.F.L. Ceramic Forum International*, vol. 70, p. 212, 1993.

[KUL 87] KULA M.-R., SCHÜTTE H., "Purification of proteins and the disruption of microbial cells", *Biotechnology Progress*, vol. 3, no. 1, p. 31, 1987.

[LAN 87a] LANDWEHR D., PAHL M.H., "Zerkleinerungsvorgänge in einer Turbomühle Teil 1", *Aufbereitungs Technik*, no. 2, p. 57, 1987.

[LAN 87b] LANDWEHR D., PAHL M.H., "Zerkleinerungsvorgänge in einer Turbomühle Teil 2", *Aufbereitungs Technik*, no. 4, p. 188, 1987.

[LEV 62] LEVENSPIEL O., *Chemical Reaction Engineering*, Wiley, New York, 1962.

[LIC 56] LICHNEROWICZ A., *Algèbre et analyse linéaire*, Masson, 1956.

[LOC 72] LOCHER F.W., SEEBACH H.M., "Influence of adsorption on industrial grinding", *Industrial and Engineering Chemistry Process Design Development*, vol. 11, no. 2, p. 190, 1972.

[LUC 72] LUCKIE P.T., AUSTIN M.S., AUSTIN L.G., "A review introduction to the solutions of the grinding equations by digital computation", *Mineral Science and Engineering*, vol. 4, p. 24, 1972.

[MAX 34] MAXSON W.L., CADENA F., BOND F.C., "Grindability of various ores", *Transactions of the American Institute of Mining and Metallurgy. Petroleum Engineering*, vol. 112, p. 130, 1934.

[MEH 89] MEHTA R.K., ADEL G.T., YOON R.H., "Liberation modeling and parameter estimation for multicomponent mineral systems", *Powder Technology*, vol. 58, p. 195, 1989.

[MEN 86] MENACHO J.M., "Some solutions for the kinetics of combined fracture and abrasion breakage", *Powder Technology*, vol. 49, p. 87, 1986.

[MEN 03] MENDE S., STENGER F., PEUKERT W. *et al.*, "Mechanical production and stabilization of submicron particles in stirred media mills", *Powder Technology*, vol. 132, p. 64, 2003.

[MIK 76] MIKA T.S., "A solution to the distributed parameter model of a continuous grinding mill at steady state", *Chemical Engineering Science*, vol. 31, p. 257, 1976.

[MOO 88] MOORE J.J., PEREZ R., GANGOPADHYAY A. *et al.*, "Factors affecting wear in tumbling mills: influence of composition and microstructure", *International Journal of Mineral Processing*, vol. 22, p. 313, 1988.

[MOR 92] MORREL S., "Prediction of grinding mill power", *Transactions of the Institution of Mining and Metallurgy Section A Mining Industry*, vol. 101, p. 25, 1992.

[OPO 84] OPOCZKY L., FARNADAY F., "Fine grinding and states of equilibrium", *Powder Technology*, vol. 39, p. 107, 1984.

[PAT 65] PATAT H.F., MEMPEL G., "Kinetik der Hartzerkleinerung 4 Teil Zur Zerkleinerung in Kugelmiihlen", *Chemie-Ingerieur-Technik*, vol. 37, p. 933, 1965.

[PET 85] PETERSON R.D., HERBST J.A., "Estimation of kinetic parameters of a grinding-liberation model", *International Journal of Mineral Processing*, vol. 14, p. 111, 1985.

[PRA 87] PRASHER C.L., *Crushing and Grinding Process Handbook*, Wiley, 1987.

[REI 65] REID K.J., "A solution to the batch grinding equation", *Chemical Engineering Science*, vol. 20, p. 953, 1965.

[RIT 67] RITTINGER P.R., *Lehrbuch der Aufbereitungskunde*, Berlin,Verlag von Ernst & Korn 1867.

[ROG 82] ROGERS R.S.C., "Closed form analytical solutions for models of closed circuit crushers", *Powder Technology*, vol. 32, p. 125, 1982.

[ROG 83] ROGERS R.S.C., "A generalized transfer parameter treatment of crushing and grinding circuit simulation", *Powder Technology*, vol. 36, p. 137, 1983.

[ROS 33] ROSIN P., RAMMLER E., "The law governing the fineness of powdered coal", *Journal of the Institute of Fuel*, vol. 7, p. 29, 1933.

[ROS 34] ROSIN P., RAMMLER E., "Gesetze des Mahlgutes", *Berichte der Deutschen Keramischen Gesellschaft*, vol. 15, p. 399, 1934.

[ROS 56] ROSE H.E., EVANS D.E., "The dynamics of the ball mill", *Proceedings of Mechanical Engineers*, vol. 170, p. 773, 1956.

[ROS 57] ROSE H.E., "A study of vibration milling on the basis of considerations of dynamical similarity", *Transactions of the Institute of Chemical Engineers*, vol. 35, p. 98, 1957.

[ROS 67] ROSE H.E., ENGLISH J.E., "Theoretical analysis of the performance of jaw crushers", *I.M.M. Transactions*, vol. 76, p. C32, 1967.

[SAD 75] SADLER III L.Y., STANLEY D.A., BROOKS D.R., "Attrition mill operating characteristics", *Powder Technology*, vol. 12, p. 19, 1975.

[SCH 40] SCHUHMANN R., "Principles of comminution, I-size distribution and surface calculations", *American Institute of Mining and Metallurgical Engineers, Technical Publication*, vol. 4, no. 1189, p. 1, 1940.

[SCH 72] SCHOENERT K., "Role of fracture physics in understanding comminution phenomena", *Transactions of the Society of Mining Engineers of A.I.M.E.*, vol. 252, p. 21, 1972.

[SCH 83] SCHÜTTE H., KRONER K.H., HUSTERT H. *et al.*, "Experiences with a 20 litre industrial bead mill for the disruption of microorganism", *Enzyme Microbial Technology*, vol. 5, p. 143, 1983.

[SCH 86] SCHALINUS H., "Zerkleinungs-und Klassierverhalten in Siebhammermühlen", *Chemie-Ingenieur-Technik*, vol. 58, no. 7, p. 604, 1986.

[SCH 88] SCHÄFER H.-V., GALLUS D., "Brechanlagen mit Doppelwellen Hammerbrechen zum Zerkleinern von Rohmaterialien für die Klinkerherstellung", *Zement-Kalk-Gips*, no. 10, p. 486, 1988.

[SET 83] SETE, "Du nouveau dans les broyeurs à air comprimé", *Informations Chimie*, no. 243, 1983.

[SHO 74] SHOJI K., AUSTIN L.G., "A model fot batch rod milling", *Powder Technology*, vol. 10, p. 29, 1974.

[SOM 72] SOMASUNDARAN P., LIN I.J., "Effect of the nature of environment on comminution processes", *Industrial and Engineering Chemistry Process Design and Development*, vol. 11, p. 321, 1972.

[SPI 92] SPIEGEL M.R., *Formules et tables de mathématiques*, Série Schaum, McGraw-Hill, 1992.

[STA 74] STANLEY G.G., "Mechanisms in the autogenous mill and their mathematical representation", *Journal of the South African Institute of Mining and Metallurgy*, vol. 75, p. 77, 1974.

[STE 83] STEHR N., SCHWEDES J., "Verfahrenstechische Untersuchung an einer Rührwerkskugelmühle", *Aufbereitungs-Technik*, no. 10, p. 597, 1983.

[STI 05] STINGER F., MEND S., SCHWEDES J. *et al.*, "Nanomilling in stirred media mills", *Chemical Engineering Science*, vol. 60, p. 4557, 2005.

[SWA 81] SWAROOP S.H.R., ABOUZEID A.Z.M., FUERSTENAU D.W., "Flow of particulate solids through tumbling mills", *Powder Technology*, vol. 28, p. 253, 1981.

[TAN 64] TANAKA T., "Axial flow in rotating cylinders", *Mineral Processing Cleveland*, vol. 5, p. 24, 1964.

[TAN 72] TANAKA T., "Scale-up theory of jet mills on basis of comminution kinetics", *Industrial and Engineering Chemistry Process Design and Development*, vol. 11, p. 238, 1972.

[TAN 89] TANGSATHIKULCHAI C., "Slurry density effects on ball milling in a laboratory ball mill", *Powder Technology*, vol. 59, p. 285, 1989.

[TSC 87a] TSCHORBADJISKI I., SCHALINUS H., SCHWEDES J., "Modellierung der Zerkleinerung in Hammermühlen Teil 1", *Aufbereitungs-Technik*, no. 10, p. 555, 1987.

[TSC 87b] TSCHORBADJISKI I., SCHALINUS H., SCHWEDES J., "Modellierung der Zerkleinerung in Hammermühlen Teil 2", *Aufbereitungs-Technik*, no. 12, p. 699, 1987.

[VOL 13] VOLTERRA V., *Leçons sur les équations intégrales et les équations integro-différentielles*, Gauthier Villars, 1913.

[WEL 88] WELLER K.R., STERNS V.J., ARTONE E. *et al.*, "Multicomponent models of grinding and classification for scale-up from continuous small or pilot scale circuits", *International Journal of Mineral Processing*, vol. 22, p. 119, 1988.

[WHI 84] WHITEN W.J., KAVERTSKY A., "Studies on scale-up of ball mills", *Mineral and Metallurgical Processing*, vol. 1, no. 1, p. 23, 1984.

Index

Printed in the United States
By Bookmasters